Also by Mario Livio

The Equation That Couldn't Be Solved: How Mathematical Genius Discovered the Language of Symmetry

The Golden Ratio: The Story of Phi, the World's Most Astonishing Number

The Accelerating Universe: Infinite Expansion, the Cosmological Constant, and the Beauty of the Cosmos

IS GOD A MATHEMATICIAN?

Mario Livio

Simon & Schuster
NEW YORK • LONDON • TORONTO • SYDNEY

Simon & Schuster
1230 Avenue of the Americas
New York, NY 10020

Permissions and acknowledgments for reprinted material are listed on pages 307 and 308.

Designed by Paul Dippolito

Manufactured in the United States of America

ISBN-13: 978-0-7432-9405-8

To Sofie

CONTENTS

PREFACE

When you work in cosmology—the study of the cosmos at large—one of the facts of life becomes the weekly letter, e-mail, or fax from someone who wants to describe to you *his* own theory of the universe (yes, they are invariably men). The biggest mistake you can make is to politely answer that you would like to learn more. This immediately results in an endless barrage of messages. So how can you prevent the assault? A particular tactic that I found to be quite effective (short of the impolite act of not answering at all) is to point out the true fact that as long as the theory is not precisely formulated in the language of mathematics, it is impossible to assess its relevance. This response stops most amateur cosmologists in their tracks. The reality is that without mathematics, modern-day cosmologists could not have progressed even one step in attempting to understand the laws of nature. Mathematics provides the solid scaffolding that holds together any theory of the universe. This may not sound so surprising until you realize that the nature of mathematics itself is not entirely clear. As the British philosopher Sir Michael Dummett once put it: "The two most abstract of the intellectual disciplines, philosophy and mathematics, give rise to the same perplexity: what are they *about*? The perplexity does not arise solely out of ignorance: even the practitioners of these subjects may find it difficult to answer the question."

In this book I humbly try to clarify both some aspects of the essence of mathematics and, in particular, the nature of the relation between mathematics and the world we observe. The book is definitely not meant to represent a comprehensive history of mathematics. Rather, I chronologically follow the evolution of some concepts that have direct implications for understanding the role of mathematics in our grasp of the cosmos.

Many people have contributed, directly and indirectly, over a long period of time, to the ideas presented in this book. I would like to thank Sir Michael Atiyah, Gia Dvali, Freeman Dyson, Hillel Gauchman, David Gross, Sir Roger Penrose, Lord Martin Rees, Raman Sundrum, Max Tegmark, Steven Weinberg, and Stephen Wolfram for very helpful exchanges. I am indebted to Dorothy Morgenstern Thomas for allowing me to use the complete text of Oscar Morgenstern's account of Kurt Gödel's experience with the U.S. Immigration and Naturalization Service. William Christens-Barry, Keith Knox, Roger Easton, and in particular Will Noel were kind enough to give me detailed explanations of their efforts to decipher the Archimedes Palimpsest. Special thanks are due to Laura Garbolino for providing me with crucial materials and rare files regarding the history of mathematics. I also thank the special collections departments of the Johns Hopkins University, the University of Chicago, and the Bibliothèque nationale de France, Paris, for finding some rare manuscripts for me.

I am grateful to Stefano Casertano for his help with difficult translations from Latin, and to Elizabeth Fraser and Jill Lagerstrom for their invaluable bibliographic and linguistic support (always with a smile).

Special thanks are due to Sharon Toolan for her professional help in the preparation of the manuscript for print, and to Ann Feild, Krista Wildt, and Stacey Benn for drawing some of the figures.

Every author should consider herself or himself fortunate to receive from their spouse the type of continuous support and patience that I have received from my wife, Sofie, during the long period of the writing of this book.

Finally, I would like to thank my agent, Susan Rabiner, without whose encouragement this book would have never happened. I am also deeply indebted to my editor, Bob Bender, for his careful reading of the manuscript and his insightful comments, to Johanna Li for her invaluable support with the production of the book, to Loretta Denner and Amy Ryan for copyediting, to Victoria Meyer and Katie Grinch for promoting the book, and to the entire production and marketing team at Simon & Schuster for all their hard work.

IS GOD A MATHEMATICIAN?

CHAPTER 1

A MYSTERY

A few years ago, I was giving a talk at Cornell University. One of my PowerPoint slides read: "Is God a mathematician?" As soon as that slide appeared, I heard a student in the front row gasp: "Oh God, I hope not!"

My rhetorical question was neither a philosophical attempt to define God for my audience nor a shrewd scheme to intimidate the math phobics. Rather, I was simply presenting a mystery with which some of the most original minds have struggled for centuries—the apparent omnipresence and omnipotent powers of mathematics. These are the type of characteristics one normally associates only with a deity. As the British physicist James Jeans (1877–1946) once put it: "The universe appears to have been designed by a pure mathematician." Mathematics appears to be almost too effective in describing and explaining not only the cosmos at large, but even some of the most chaotic of human enterprises.

Whether physicists are attempting to formulate theories of the universe, stock market analysts are scratching their heads to predict the next market crash, neurobiologists are constructing models of brain function, or military intelligence statisticians are trying to optimize resource allocation, they are all using mathematics. Furthermore, even though they may be applying formalisms developed in different branches of mathematics, they are still referring to the same global, coherent mathematics. What is it that gives mathematics such incredible powers? Or, as Einstein once wondered: "How is it possible that mathematics, a product of human thought that is *independent of experience* [the emphasis is mine], fits so excellently the objects of physical reality?"

This sense of utter bewilderment is not new. Some of the philosophers in ancient Greece, Pythagoras and Plato in particular, were already in awe of the apparent ability of mathematics to shape and guide the universe, while existing, as it seemed, above the powers of humans to alter, direct, or influence it. The English political philosopher Thomas Hobbes (1588–1679) could not hide his admiration either. In *Leviathan,* Hobbes's impressive exposition of what he regarded as the foundation of society and government, he singled out geometry as the paradigm of rational argument:

> Seeing then that truth consisteth in the right ordering of names in our affirmations, a man that seeketh precise truth had need to remember what every name he uses stands for, and to place it accordingly; or else he will find himself entangled in words, as a bird in lime twigs; the more he struggles, the more belimed. And therefore in geometry (which is the only science that it hath pleased God hitherto to bestow on mankind), men begin at settling the significations of their words; which settling of significations, they call definitions, and place them in the beginning of their reckoning.

Millennia of impressive mathematical research and erudite philosophical speculation have done relatively little to shed light on the enigma of the power of mathematics. If anything, the mystery has in some sense even deepened. Renowned Oxford mathematical physicist Roger Penrose, for instance, now perceives not just a single, but a triple mystery. Penrose identifies three different "worlds": the *world of our conscious perceptions*, the *physical world,* and the *Platonic world of mathematical forms.* The first world is the home of all of our mental images—how we perceive the faces of our children, how we enjoy a breathtaking sunset, or how we react to the horrifying images of war. This is also the world that contains love, jealousy, and prejudices, as well as our perception of music, of the smells of food, and of fear. The second world is the one we normally refer to as physical reality. Real flowers, aspirin tablets, white clouds, and jet airplanes reside in this world, as do galaxies, planets, atoms, baboon hearts, and human brains. The Platonic world of mathematical forms, which to Penrose has an actual reality comparable to that of the physical and the mental worlds, is the motherland of mathematics. This is where you will find the natural numbers 1, 2, 3, 4, . . . , all the shapes and theorems of Euclidean geometry, Newton's laws of motion, string theory, catastrophe theory, and mathematical models of stock market behavior. And now, Penrose observes, come the three mysteries. First, the world of physical reality seems to obey laws that actually reside in the world of mathematical forms. This was the puzzle that left Einstein perplexed. Physics Nobel laureate Eugene Wigner (1902–95) was equally dumbfounded:

> The miracle of the appropriateness of the language of mathematics to the formulation of the laws of physics is a wonderful gift which we neither understand nor deserve. We should be grateful for it and hope that it will remain valid in future research and that it will extend, for better or worse, to our pleasure, even though perhaps also to our bafflement, to wide branches of learning.

Second, the perceiving minds themselves—the dwelling of our conscious perceptions—somehow managed to emerge from the physical world. How was *mind* literally born out of *matter*? Would we ever be able to formulate a theory of the workings of consciousness that would be as coherent and as convincing as, say, our current theory of electromagnetism? Finally, the circle is mysteriously closed. Those perceiving minds were miraculously able to gain access to the mathematical world by discovering or creating and articulating a treasury of abstract mathematical forms and concepts.

Penrose does not offer an explanation for any of the three mysteries. Rather, he laconically concludes: "No doubt there are not really three worlds but *one,* the true nature of which we do not even glimpse at present." This is a much more humble admission than the response of the schoolmaster in the play *Forty Years On* (written by the English author Alan Bennett) to a somewhat similar question:

Foster: I'm still a bit hazy about the Trinity, sir.

Schoolmaster: Three in one, one in three, perfectly straightforward. Any doubts about that see your maths master.

The puzzle is even more entangled than I have just indicated. There are actually two sides to the success of mathematics in explaining the world around us (a success that Wigner dubbed "the unreasonable effectiveness of mathematics"), one more astonishing than the other. First, there is an aspect one might call "active." When physicists wander through nature's labyrinth, they light their way by mathematics—the tools they use and develop, the models they construct, and the explanations they conjure are all mathematical in nature. This, on the face of it, is a miracle in itself. Newton observed a falling apple, the Moon, and tides on the beaches (I'm not even sure if he ever saw those!), not mathematical equations. Yet he was somehow able to extract from all of these natural phenomena, clear, concise, and unbelievably accurate mathematical laws of nature. Similarly, when the Scottish physicist James Clerk Maxwell (1831–79) extended the framework of classical physics to include *all* the electric and magnetic phenomena that were known in the 1860s, he did so by means of just four mathematical equations. Think about this for a moment. The explanation of a collection of experimental results in electromagnetism and light, which had previously taken volumes to describe, was reduced to four succinct equations. Einstein's general relativity is even more astounding—it is a perfect example of an extraordinarily precise, self-consistent mathematical theory of something as fundamental as the structure of space and time.

But there is also a "passive" side to the mysterious effectiveness of mathematics, and it is so surprising that the "active" aspect pales by comparison. Concepts and relations explored by mathematicians only for pure reasons—with absolutely no application in mind—turn out decades (or sometimes centuries) later to be the unexpected solutions to problems grounded in physical reality! How is that possible? Take for instance the somewhat amusing case of the eccentric British mathematician Godfrey Harold Hardy (1877–1947). Hardy was so proud of the fact that his work consisted of nothing but pure mathematics that he emphatically declared: "No discovery of mine has made, or is likely to make, directly or indirectly, for good or ill, the least difference to the amenity of the world." Guess what—he was wrong. One of his works was reincarnated as the Hardy-Weinberg law (named after Hardy and the German physician Wilhelm Weinberg [1862–1937]), a fundamental principle used by geneticists to study the evolution of populations. Put simply, the Hardy-Weinberg law states that if a large population is mating totally at random (and migration, mutation, and selection do not occur), then the genetic constitution remains constant from one generation to the next. Even Hardy's seemingly abstract work on *number theory*—the study of the properties of the natural numbers—found unexpected applications. In 1973, the British mathematician Clifford Cocks used the theory of numbers to create a breakthrough in cryptography—the development of codes. Cocks's discovery made another statement by Hardy obsolete. In his famous book *A Mathematician's Apology,* published in 1940, Hardy pronounced: "No one has yet discovered any war-like purpose to be served by the theory of numbers." Clearly, Hardy was yet again in error. Codes have been absolutely essential for military communications. So even Hardy, one of the most vocal critics of applied mathematics, was "dragged" (probably kicking and screaming, if he had been alive) into producing useful mathematical theories.

But this is only the tip of the iceberg. Kepler and Newton discovered that the planets in our solar system follow orbits in the shape of ellipses—the very curves studied by the Greek mathematician Menaechmus (fl. ca. 350 BC) two millennia earlier. The new types of geometries outlined by Georg Friedrich Bernhard Riemann (1826–66) in a classic lecture in 1854 turned out to be precisely the tools that Einstein needed to explain the cosmic fabric. A mathematical "language" called group theory, developed by the young prodigy Évariste Galois (1811–32) simply to determine the solvability of algebraic equations, has today become the language used by physicists, engineers, linguists, and even anthropologists to describe all the symmetries of the world. Moreover, the concept of mathematical symmetry patterns has, in some sense, turned the entire scientific process on its head. For centuries the route to understanding the workings of the cosmos started with a collection of experimental or observational facts, from which, by trial and error, scientists attempted to formulate general laws of nature. The scheme was to begin with local observations and build the jigsaw puzzle piece by piece. With the recognition in the twentieth century that well-defined mathematical designs underlie the structure of the subatomic world, modern-day physicists started to do precisely the opposite. They put the mathematical symmetry principles *first*, insisting that the laws of nature and indeed the basic building blocks of matter should follow certain patterns, and they deduced the general laws from these requirements. How does nature know to obey these abstract mathematical symmetries?

In 1975, Mitch Feigenbaum, then a young mathematical physicist at Los Alamos National Laboratory, was playing with his HP-65 pocket calculator. He was examining the behavior of a simple equation. He noticed that a sequence of numbers that appeared in the calculations was getting closer and closer to a particular number: 4.669 . . . To his amazement, when he examined other equations, the same curious number appeared again. Feigenbaum soon concluded that his discovery represented something universal, which somehow marked the transition from order to chaos, even though he had no explanation for it. Not surprisingly, physicists were very skeptical at first. After all, why should the same number characterize the behavior of what appeared to be rather different systems? After six months of professional refereeing, Feigenbaum's first paper on the topic was rejected. Not much later, however, experiments showed that when liquid helium is heated from below it behaves precisely as predicted by Feigenbaum's universal solution. And this was not the only system found to act this way. Feigenbaum's astonishing number showed up in the transition from the orderly flow of a fluid to turbulence, and even in the behavior of water dripping from a tap.

The list of such "anticipations" by mathematicians of the needs of various disciplines of later generations just goes on and on. One of the most fascinating examples of the mysterious and unexpected interplay between mathematics and the real (physical) world is provided by the story of *knot theory*—the mathematical study of knots. A mathematical knot resembles an ordinary knot in a string, with the string's ends spliced together. That is, a mathematical knot is a closed curve with no loose ends. Oddly, the main impetus for the development of mathematical knot theory came from an incorrect model for the atom that was developed in the nineteenth century. Once that model was abandoned—only two decades after its conception—knot theory continued to evolve as a relatively obscure branch of pure mathematics. Amazingly, this abstract endeavor suddenly found extensive modern applications in topics ranging from the molecular structure of DNA to string theory—the attempt to unify the subatomic world with gravity. I shall return to this remarkable tale in chapter 8, because its circular history is perhaps the best demonstration of how branches of mathematics can emerge from attempts to explain physical reality, then how they wander into the abstract realm of mathematics, only to eventually return unexpectedly to their ancestral origins.

Discovered or Invented?

Even the brief description I have presented so far already provides overwhelming evidence of a universe that is either governed by mathematics or, at the very least, susceptible to analysis through mathematics. As this book will show, much, and perhaps all, of the human enterprise also seems to emerge from an underlying mathematical facility, even where least expected. Examine, for instance, an example from the world of finance—the Black-Scholes option pricing formula (1973). The Black-Scholes model won its originators (Myron Scholes and Robert Carhart Merton; Fischer Black passed away before the prize was awarded) the Nobel Memorial Prize in economics. The key equation in the model enables the understanding of stock option pricing (options are financial instruments that allow bidders to buy or sell stocks at a future point in time, at agreed-upon prices). Here, however, comes a surprising fact. At the heart of this model lies a phenomenon that had been studied by physicists for decades—Brownian motion, the state of agitated motion exhibited by tiny particles such as pollen suspended in water or smoke particles in the air. Then, as if that were not enough, the same equation also applies to the motion of hundreds of thousands of stars in star clusters. Isn't this, in the language of *Alice in Wonderland,* "curiouser and curiouser"? After all, whatever the cosmos may be doing, business and finance are definitely worlds created by the human mind.

Or, take a common problem encountered by electronic board manufacturers and designers of computers. They use laser drills to make tens of thousands of holes in their boards. In order to minimize the cost, the computer designers do not want their drills to behave as "accidental tourists." Rather, the problem is to find the shortest "tour" among the holes, that visits each hole position exactly once. As it turns out, mathematicians have investigated this exact problem, known as the *traveling salesman problem*, since the 1920s. Basically, if a salesperson or a politician on the campaign trail needs to travel in the most economical way to a given number of cities, and the cost of travel between each pair of cities is known, then the traveler must somehow figure out the cheapest way of visiting all the cities and returning to his or her starting point. The traveling salesman problem was solved for 49 cities in the United States in 1954. By 2004, it was solved for 24,978 towns in Sweden. In other words, the electronics industry, companies routing trucks for parcel pickups, and even Japanese manufacturers of pinball-like pachinko machines (which have to hammer thousands of nails) have to rely on mathematics for something as simple as drilling, scheduling, or the physical design of computers.

Mathematics has even penetrated into areas not traditionally associated with the exact sciences. For instance, there is a *Journal of Mathematical Sociology* (which in 2006 was in its thirtieth volume) that is oriented toward a mathematical understanding of complex social structures, organizations, and informal groups. The journal articles address topics ranging from a mathematical model for predicting public opinion to one predicting interaction in social groups.

Going in the other direction—from mathematics into the humanities—the field of computational linguistics, which originally involved only computer scientists, has now become an interdisciplinary research effort that brings together linguists, cognitive psychologists, logicians, and artificial intelligence experts, to study the intricacies of languages that have evolved naturally.

Is this some mischievous trick played on us, such that all the human struggles to grasp and comprehend ultimately lead to uncovering the more and more subtle fields of mathematics upon which the universe and we, its complex creatures, were all created? Is mathematics, as educators like to say, the hidden textbook—the one the professor teaches from—while giving his or her students a much lesser version so that he or she will seem all the wiser? Or, to use the biblical metaphor, is mathematics in some sense the ultimate fruit of the tree of knowledge?

As I noted briefly at the beginning of this chapter, the unreasonable effectiveness of mathematics creates many intriguing puzzles: Does mathematics have an existence that is entirely independent of the human mind? In other words, are we merely *discovering* mathematical verities, just as astronomers discover previously unknown galaxies? Or, is mathematics nothing but a human *invention*? If mathematics indeed exists in some abstract fairyland, what is the relation between this mystical world and physical reality? How does the human brain, with its known limitations, gain access to such an immutable world, outside of space and time? On the other hand, if mathematics is merely a human invention and it has no existence outside our minds, how can we explain the fact that the invention of so many mathematical truths miraculously anticipated questions about the cosmos and human life not even posed until many centuries later? These are not easy questions. As I will show abundantly in this book, even modern-day mathematicians, cognitive scientists, and philosophers don't agree on the answers. In 1989, the French mathematician Alain Connes, winner of two of the most prestigious prizes in mathematics, the Fields Medal (1982) and the Crafoord Prize (2001), expressed his views very clearly:

> Take prime numbers [those divisible only by one and themselves], for example, which as far as I'm concerned, constitute a more stable reality than the material reality that surrounds us. The working mathematician can be likened to an explorer who sets out to discover the world. One discovers basic facts from experience. In doing simple calculations, for example, one real-

> izes that the series of prime numbers seems to go on without end. The mathematician's job, then, is to demonstrate that there exists an infinity of prime numbers. This is, of course, an old result due to Euclid. One of the most interesting consequences of this proof is that if someone claims one day to have found the greatest prime number, it will be easy to show that he's wrong. The same is true for any proof. We run up therefore against a reality every bit as incontestable as physical reality.

Martin Gardner, the famous author of numerous texts in recreational mathematics, also takes the side of mathematics as a *discovery*. To him, there is no question that numbers and mathematics have their own existence, whether humans know about them or not. He once wittily remarked: "If two dinosaurs joined two other dinosaurs in a clearing, there would be four there, even though no humans were around to observe it, and the beasts were too stupid to know it." As Connes emphasized, supporters of the "mathematics-as-a-discovery" perspective (which, as we shall see, conforms with the Platonic view) point out that once any particular mathematical concept has been grasped, say the natural numbers 1, 2, 3, 4, . . . , then we are up against undeniable facts, such as $3^2 + 4^2 = 5^2$, irrespective of what we think about these relations. This gives at least the impression that we are in contact with an existing reality.

Others disagree. While reviewing a book in which Connes presented his ideas, the British mathematician Sir Michael Atiyah (who won the Fields Medal in 1966 and the Abel Prize in 2004) remarked:

> Any mathematician must sympathize with Connes. We all feel that the integers, or circles, really exist in some abstract sense and the Platonic view [which will be described in detail in chapter 2] is extremely seductive. But can we really defend it? Had the universe been one dimensional or even discrete it is difficult to see how geometry could have evolved. It might seem that with the integers we are on firmer ground, and that counting is really a primordial notion. But let us imagine that intelligence had resided, not in mankind, but in some vast

> solitary and isolated jelly-fish, buried deep in the depths of the Pacific Ocean. It would have no experience of individual objects, only with the surrounding water. Motion, temperature and pressure would provide its basic sensory data. In such a pure continuum the discrete would not arise and there would be nothing to count.

Atiyah therefore believes that "man has *created* [the emphasis is mine] mathematics by idealizing and abstracting elements of the physical world." Linguist George Lakoff and psychologist Rafael Núñez agree. In their book *Where Mathematics Comes From,* they conclude: "Mathematics is a natural part of being human. It arises from our bodies, our brains, and our everyday experiences in the world."

The viewpoint of Atiyah, Lakoff, and Núñez raises another interesting question. If mathematics is entirely a human invention, is it truly universal? In other words, if extraterrestrial intelligent civilizations exist, would they invent the same mathematics? Carl Sagan (1934–96) used to think that the answer to the last question was in the affirmative. In his book *Cosmos,* when he discussed what type of signal an intelligent civilization would transmit into space, he said: "It is extremely unlikely that any natural physical process could transmit radio messages containing prime numbers only. If we received such a message we would deduce a civilization out there that was at least fond of prime numbers." But how certain is that? In his recent book *A New Kind of Science,* mathematical physicist Stephen Wolfram argued that what we call "our mathematics" may represent just one possibility out of a rich variety of "flavors" of mathematics. For instance, instead of using rules based on mathematical equations to describe nature, we could use different types of rules, embodied in simple computer programs. Furthermore, some cosmologists have recently discussed even the possibility that our universe is but one member of a *multiverse*—a huge ensemble of universes. If such a multiverse indeed exists, would we really expect the other universes to have the same mathematics?

Molecular biologists and cognitive scientists bring to the table yet another perspective, based on studies of the faculties of the brain. To some of these researchers, mathematics is not very different from language. In other words, in this "cognitive" scenario, after eons during which humans stared at two hands, two eyes, and two breasts, an abstract definition of the number 2 has emerged, much in the same way that the word "bird" has come to represent many two-winged animals that can fly. In the words of the French neuroscientist Jean-Pierre Changeux: "For me the axiomatic method [used, for instance, in Euclidean geometry] is the expression of cerebral faculties connected with the use of the human brain. For what characterizes language is precisely its generative character." But, if mathematics is just another language, how can we explain the fact that while children study languages easily, many of them find it so hard to study mathematics? The Scottish child prodigy Marjory Fleming (1803–11) charmingly described the type of difficulties students encounter with mathematics. Fleming, who never lived to see her ninth birthday, left journals that comprise more than nine thousand words of prose and five hundred lines of verse. In one place she complains: "I am now going to tell you the horrible and wretched plague that my multiplication table gives me; you can't conceive it. The most devilish thing is 8 times 8 and 7 times 7; it is what nature itself can't endure."

A few of the elements in the intricate questions I have presented can be recast into a different form: Is there any difference in basic kind between mathematics and other expressions of the human mind, such as the visual arts or music? If there isn't, why does mathematics exhibit an imposing coherence and self-consistency that does not appear to exist in any other human creation? Euclid's geometry, for instance, remains as correct today (where it applies) as it was in 300 BC; it represents "truths" that are forced upon us. By contrast, we are neither compelled today to listen to the same music the ancient Greeks listened to nor to adhere to Aristotle's naïve model of the cosmos.

Very few scientific subjects today still make use of ideas that can be three thousand years old. On the other hand, the latest research in mathematics may refer to theorems that were published last year, or last week, but it may also use the formula for the surface area of a sphere proved by Archimedes around 250 BC! The nineteenth century knot model of the atom survived for barely two decades because new discoveries proved elements of the theory to be in error. This is how science progresses. Newton gave credit (or not! see chapter 4) for his great vision to those giants upon whose shoulders he stood. He might also have apologized to those giants whose work he had made obsolete.

This is not the pattern in mathematics. Even though the formalism needed to prove certain results might have changed, the mathematical results themselves do not change. In fact, as mathematician and author Ian Stewart once put it, "There is a word in mathematics for previous results that are later changed—they are simply called *mistakes.*" And such mistakes are judged to be mistakes not because of new findings, as in the other sciences, but because of a more careful and rigorous reference to the same old mathematical truths. Does this indeed make mathematics God's native tongue?

If you think that understanding whether mathematics was invented or discovered is not that important, consider how loaded the difference between "invented" and "discovered" becomes in the question: Was God invented or discovered? Or even more provocatively: Did God create humans in his own image, or did humans invent God in their own image?

I will attempt to tackle many of these intriguing questions (and quite a few additional ones) and their tantalizing answers in this book. In the process, I shall review insights gained from the works of some of the greatest mathematicians, physicists, philosophers, cognitive scientists, and linguists of past and present centuries. I shall also seek the opinions, caveats, and reservations of many modern thinkers. We start this exciting journey with the groundbreaking perspective of some of the very early philosophers.

CHAPTER 2

MYSTICS: THE NUMEROLOGIST AND THE PHILOSOPHER

Humans have always been driven by a desire to understand the cosmos. Their efforts to get to the bottom of "What does it all mean?" far exceeded those needed for mere survival, improvement in the economic situation, or the quality of life. This does not mean that everybody has always actively engaged in the search for some natural or metaphysical order. Individuals struggling to make ends meet can rarely afford the luxury of contemplating the meaning of life. In the gallery of those who hunted for patterns underlying the universe's perceived complexity, a few stood head and shoulders above the rest.

To many, the name of the French mathematician, scientist, and philosopher René Descartes (1596–1650) is synonymous with the birth of the modern age in the philosophy of science. Descartes was one of the principal architects of the shift from a description of the natural world in terms of properties directly perceived by our senses to explanations expressed through mathematically well-defined quantities. Instead of vaguely characterized feelings, smells, colors, and sensations, Descartes wanted scientific explanations to probe to the very fundamental microlevel, and to use the language of mathematics:

> I recognize no matter in corporeal things apart from that which the geometers call *quantity,* and take as the object of their demonstrations . . . And since all natural phenomena can be explained in this way, I do not think that any other principles are either admissible or desirable in physics.

Interestingly, Descartes excluded from his grand scientific vision the realms of "thought and mind," which he regarded as independent of the mathematically explicable world of matter. While there is no doubt that Descartes was one of the most influential thinkers of the past four centuries (and I shall return to him in chapter 4), he was not the first to have exalted mathematics to a central position. Believe it or not, sweeping ideas of a cosmos permeated and governed by mathematics—ideas that in some sense went even further than those of Descartes—had first been expressed, albeit with a strong mystical flavor, more than two millennia earlier. The person to whom legend ascribes the perception that the human soul is "at music" when engaged in pure mathematics was the enigmatic Pythagoras.

Pythagoras

Pythagoras (ca. 572–497 BC) may have been the first person who was both an influential natural philosopher and a charismatic spiritual philosopher—a scientist and a religious thinker. In fact, he is credited with introducing the words "philosophy," meaning love of wisdom, and "mathematics"—the learned disciplines. Even though none of Pythagoras's own writings have survived (if these writings ever existed, since much was communicated orally), we do have three detailed, if only partially reliable, biographies of Pythagoras from the third century. A fourth, anonymous one was preserved in the writings of the Byzantine patriarch and philosopher Photius (ca. AD 820–91). The main problem with attempting to assess Pythagoras's personal contributions lies in the fact that his followers and disciples—the Pythagoreans—invariably attribute all their ideas to him. Consequently, even Aristotle (384–322 BC) finds it difficult to identify which portions of the Pythagorean philosophy can safely be ascribed to Pythagoras himself, and he generally refers to "the Pythagoreans" or "the so-called Pythagoreans." Nevertheless, given Pythagoras's fame in later tradition, it is generally assumed that he was the originator of at least some of the Pythagorean theories to which Plato and even Copernicus felt indebted.

There is little doubt that Pythagoras was born in the early sixth century BC on the island of Samos, just off the coast of modern-day Turkey. He may have traveled extensively early in life, especially to Egypt and perhaps Babylon, where he would have received at least part of his mathematical education. Eventually he emigrated to a small Greek colony in Croton, near the southern tip of Italy, where an enthusiastic group of students and followers quickly gathered around him.

The Greek historian Herodotus (ca. 485–425 BC) referred to Pythagoras as "the most able philosopher among the Greeks," and the pre-Socratic philosopher and poet Empedocles (ca. 492–432 BC) added in admiration: "But there was among them a man of prodigious knowledge, who acquired the profoundest wealth of understanding and was the greatest master of skilled arts of every kind; for whenever he willed with all his heart, he could with ease discern each and every truth in his ten—nay, twenty men's lives." Still, not all were equally impressed. In comments that appear to stem from some personal rivalry, the philosopher Heraclitus of Ephesus (ca. 535–475 BC) acknowledges Pythagoras's broad knowledge, but he is also quick to add disparagingly: "Much learning does not teach wisdom; otherwise it would have taught Hesiod [a Greek poet who lived around 700 BC] and Pythagoras."

Pythagoras and the early Pythagoreans were neither mathematicians nor scientists in the strict sense of these terms. Rather, a metaphysical philosophy of the meaning of numbers lay at the heart of their doctrines. To the Pythagoreans, numbers were both living entities and universal principles, permeating everything from the heavens to human ethics. In other words, numbers had two distinct, complementary aspects. On one hand, they had a tangible physical existence; on the other, they were abstract prescriptions on which everything was founded. For instance, the *monad* (the number 1) was understood both as the generator of all other numbers, an entity as real as water, air, and fire that participated in the structure of the physical world, and as an idea—the metaphysical unity at the source of all creation. The English historian of philosophy Thomas Stanley (1625–78) described beautifully (if in seventeenth century English) the two meanings that the Pythagoreans associated with numbers:

> Number is of two kinds the Intellectual (or immaterial) and the Sciential. The Intellectual is that eternal substance of Number, which Pythagoras in his Discourse concerning the Gods asserted to be the *principle most providential of all Heaven and Earth, and the nature that is betwixt them* . . . This is that which is termed *the principle, fountain, and root of all things* . . . Sciential Number is that which Pythagoras defines as *the extension and production into act of the seminal reasons which are in the Monad, or a heap of Monads.*

So numbers were not simply tools to denote quantities or amounts. Rather, numbers had to be discovered, and they were the formative agents that are active in nature. Everything in the universe, from material objects such as the Earth to abstract concepts such as justice, was number through and through.

The fact that someone would find numbers fascinating is perhaps not surprising in itself. After all, even the ordinary numbers encountered in everyday life have interesting properties. Take the number of days in a year—365. You can easily check that 365 is equal to the sums of three consecutive squares: $365 = 10^2 + 11^2 + 12^2$. But this is not all; it is also equal to the sum of the next two squares ($365 = 13^2 + 14^2$)! Or, examine the number of days in the lunar month—28. This number is the sum of all of its divisors (the numbers that divide it with no remainder): $28 = 1 + 2 + 4 + 7 + 14$. Numbers with this special property are called *perfect numbers* (the first four perfect numbers are 6, 28, 496, 8218). Note also that 28 is the sum of the cubes of the first two odd numbers: $28 = 1^3 + 3^3$. Even a number as widely used in our decimal system as 100 has its own peculiarities: $100 = 1^3 + 2^3 + 3^3 + 4^3$.

OK, so numbers can be intriguing. Still, one may wonder what was the origin of the Pythagorean doctrine of numbers? How did the idea arise that not only do all things possess number, but that all things *are* numbers? Since Pythagoras either wrote nothing down or his writings have been destroyed, it is not easy to answer this question. The surviving impression of Pythagoras's reasoning is based on a small number of pre-Platonic fragments and on much later, less reliable discussions, mostly by Platonic and Aristotelian philosophers. The picture that emerges from assembling the different clues suggests that the explanation of the obsession with numbers may be found in the preoccupation of the Pythagoreans with two apparently unrelated activities: experiments in music and observations of the heavens.

To understand how those mysterious connections among numbers, the heavens, and music materialized, we have to start from the interesting observation that the Pythagoreans had a way of *figuring* numbers by means of pebbles or dots. For instance, they arranged the natural numbers 1, 2, 3, 4, . . . as collections of pebbles to form triangles (as in figure 1). In particular, the triangle constructed out of the first four integers (arranged in a triangle of ten pebbles) was called the *Tetraktys* (meaning quaternary, or "fourness"), and was taken by the Pythagoreans to symbolize perfection and the elements that comprise it. This fact was documented in a story about Pythagoras by the Greek satirical author Lucian (ca. AD 120–80). Pythagoras asks someone to count. As the man counts "1, 2, 3, 4," Pythagoras interrupts him, "Do you see? What you take for 4 is 10, a perfect triangle and our oath." The Neoplatonic philosopher Iamblichus (ca. AD 250–325) tells us that the oath of the Pythagoreans was indeed:

Figure 1

I swear by the discoverer of the Tetraktys,
Which is the spring of all our wisdom,
The perennial root of Nature's fount.

Why was the Tetraktys so revered? Because to the eyes of the sixth century BC Pythagoreans, it seemed to outline the entire nature of the universe. In geometry—the springboard to the Greeks' epochal revolution in thought—the number 1 represented a point •, 2 represented a line •—•, 3 represented a surface △, and 4 represented a three-dimensional tetrahedral solid. The Tetraktys therefore appeared to encompass all the perceived dimensions of space.

But that was only the beginning. The Tetraktys made an unexpected appearance even in the scientific approach to music. Pythagoras and the Pythagoreans are generally credited with the discovery that dividing a string by simple consecutive integers produces harmonious and consonant intervals—a fact figuring in any performance by a string quartet. When two similar strings are plucked simultaneously, the resulting sound is pleasing when the lengths of the strings are in simple proportions. For instance, strings of equal length (1:1 ratio) produce a unison; a ratio of 1:2 produces the octave; 2:3 gives the perfect fifth; and 3:4 the perfect fourth. In addition to its all-embracing spatial attributes, therefore, the Tetraktys could also be seen as representing the mathematical ratios that underlie the harmony of the musical scale. This apparently magical union of space and music generated for the Pythagoreans a powerful symbol and gave them a feeling of *harmonia* ("fitting together") of the *kosmos* ("the beautiful order of things").

And where do the heavens fit into all of this? Pythagoras and the Pythagoreans played a role in the history of astronomy that, while not critical, was not negligible either. They were among the first to maintain that the Earth was spherical in form (probably because of the perceived mathematico-aesthetic superiority of the sphere). They were also probably the first to state that the planets, the Sun, and the Moon have an independent motion of their own from west to east, in a direction opposite to the daily (apparent) rotation of the sphere of the fixed stars. These enthusiastic observers of the midnight sky could not have missed the most obvious properties of the stellar constellations—shape and number. Each constellation is recognized by the number of stars that compose it and by the geometrical figure that these stars form. But these two characteristics were precisely the essential ingredients of the Pythagorean doctrine of numbers, as exemplified by the Tetraktys. The Pythagoreans were so enraptured by the dependency of geometrical figures, stellar constellations, and musical harmonies on numbers that numbers became both the building blocks from which the universe was constructed and the principles behind its existence. No wonder then that Pythagoras's maxim was stated emphatically as "All things accord in number."

We can find a testament to how seriously the Pythagoreans took this maxim in two of Aristotle's remarks. In one place in his collected treatise *Metaphysics* he says: "The so-called Pythagoreans applied themselves to mathematics, and were the first to develop this science; and through studying it they came to believe that its principles are the principles of everything." In another passage, Aristotle vividly describes the veneration of numbers and the special role of the Tetraktys: "Eurytus [a pupil of the Pythagorean Philolaus] settled what is the number of what object (e.g., this is the number of a man, that of a horse) and imitated the shapes of living things by pebbles *after the manner of those who bring numbers into the form of triangle or square.*" The last sentence ("the form of triangle or square") alludes both to the Tetraktys and to yet another fascinating Pythagorean construction—the gnomon.

The word "gnomon" (a "marker") originates from the name of a Babylonian astronomical time-measurement device, similar to a sundial. This apparatus was apparently introduced into Greece by Pythagoras's teacher—the natural philosopher Anaximander (ca. 611–547 BC). There can be no doubt that the pupil was influenced by his tutor's ideas in geometry and their application to *cosmology*—the study of the universe as a whole. Later, the term "gnomon" was used for an instrument for drawing right angles, similar to a carpenter's square, or for the right-angled figure that, when added to a square, makes up a larger square (as in figure 2). Note that if you add, say, to a 3×3 square, seven pebbles in a shape that forms a right angle (a

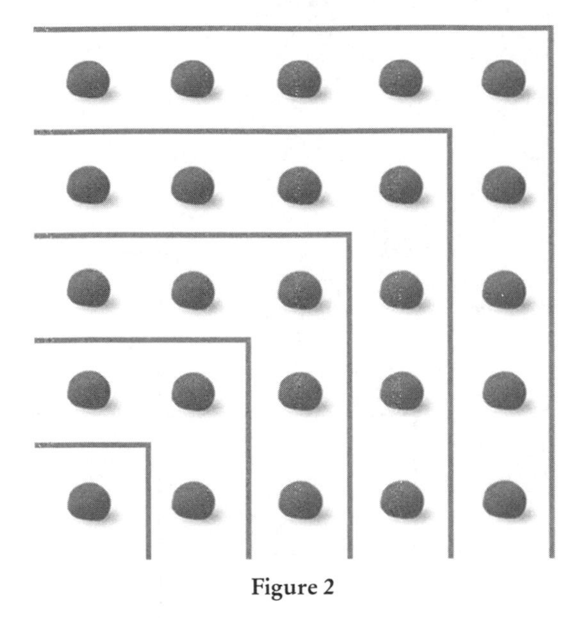

Figure 2

gnomon), you obtain a square composed of sixteen (4×4) pebbles. This is a figurative representation of the following property: In the sequence of odd integers 1, 3, 5, 7, 9, . . . , the sum of any number of successive members (starting from 1) always forms a square number. For instance, $1 = 1^2$; $1 + 3 = 4 = 2^2$; $1 + 3 + 5 = 9 = 3^2$; $1 + 3 + 5 + 7 = 16 = 4^2$; $1 + 3 + 5 + 7 + 9 = 25 = 5^2$, and so on. The Pythagoreans regarded this intimate relation between the gnomon and the square that it "embraces" as a symbol of knowledge in general, where the knowing is "hugging" the known. Numbers were therefore not limited to a description of the physical world, but were supposed to be at the root of mental and emotional processes as well.

The square numbers associated with the gnomons may have also been precursors to the famous *Pythagorean theorem.* This celebrated mathematical statement holds that for any right triangle (figure 3), a square drawn on the hypotenuse is equal in area to the sum of the squares drawn on the sides. The discovery of the theorem was "documented" humorously in a famous *Frank and Ernest* cartoon (figure 4). As the gnomon in figure 2 shows, adding a square gnomon number, $9 = 3^2$, to a 4×4 square makes a new, 5×5 square: $3^2 + 4^2 = 5^2$. The numbers 3, 4, 5 can therefore represent the lengths of the sides of

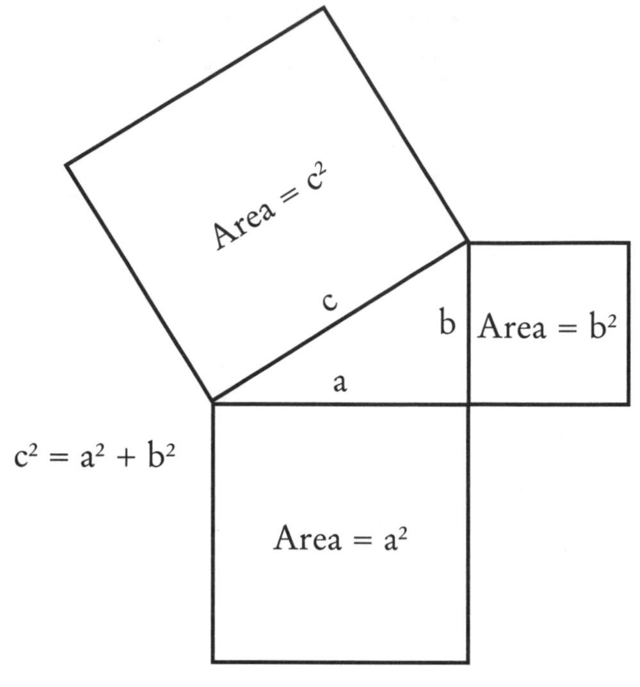

Figure 3

Figure 4

a right triangle. Integer numbers that have this property (e.g., 5, 12, 13; since $5^2 + 12^2 = 13^2$) are called "Pythagorean triples."

Few mathematical theorems enjoy the same "name recognition" as Pythagoras's. In 1971, when the Republic of Nicaragua selected the "ten mathematical equations that changed the face of the earth" as a theme for a set of stamps, the Pythagorean theorem appeared on the second stamp (figure 5; the first stamp depicted "$1 + 1 = 2$").

Was Pythagoras truly the first person to have formulated the well-known theorem attributed to him? Some of the early Greek historians certainly thought so. In a commentary on *The Elements*—the massive treatise on geometry and theory of numbers written by Euclid (ca. 325–265 BC)—the Greek philosopher Proclus (ca. AD 411–85) wrote: "If we listen to those who wish to recount ancient history, we may find some who refer this theorem to Pythagoras, and say that he sacrificed an ox in honor of the discovery." However, Pythagorean triples can already be found in the Babylonian cuneiform tablet known as Plimton 322, which dates back roughly to the time of the dynasty of Hammurabi (ca. 1900–1600 BC). Furthermore, geometrical constructions based on the Pythagorean theorem were found in India,

Figure 5

in relation to the building of altars. These constructions were clearly known to the author of the Satapatha Brahmana (the commentary on ancient Indian scriptural texts), which was probably written at least a few hundred years before Pythagoras. But whether Pythagoras was the originator of the theorem or not, there is no doubt that the recurring connections that were found to weave numbers, shapes, and the universe together took the Pythagoreans one step closer to a detailed metaphysic of order.

Another idea that played a central role in the Pythagorean world was that of *cosmic opposites.* Since the pattern of opposites was the underlying principle of the early Ionian scientific tradition, it was only natural for the order-obsessed Pythagoreans to adopt it. In fact, Aristotle tells us that even a medical doctor named Alcmaeon, who lived in Croton at the same time that Pythagoras had his famous school there, subscribed to the notion that all things are balanced in pairs. The principal pair of opposites consisted of the *limit,* represented by the odd numbers, and the *unlimited,* represented by the even. The limit was the force that introduces order and harmony into the wild, unbridled unlimited. Both the complexities of the universe at large and the intricacies of human life, microcosmically, were thought to consist of and be directed by a series of opposites that somehow fit together. This rather black-and-white vision of the world was summarized in a "table of opposites" that was preserved in Aristotle's *Metaphysics:*

Limit	Unlimited
Odd	Even
One	Plurality
Right	Left
Male	Female
Rest	Motion
Straight	Curved
Light	Darkness
Good	Evil
Square	Oblong

The basic philosophy expressed by the table of opposites was not confined to ancient Greece. The Chinese yin and yang, with the yin representing negativity and darkness and the yang the bright principle, depict the same picture. Sentiments that are not too different were carried over into Christianity, through the concepts of heaven and hell (and even into American presidential statements such as "You are either with us, or you are with the terrorists"). More generally, it has always been true that the meaning of life has been illuminated by death, and of knowledge by comparing it to ignorance.

Not all the Pythagorean teachings had to do directly with numbers. The lifestyle of the tightly knit Pythagorean society was also based on vegetarianism, a strong belief in metempsychosis—the immortality and transmigration of souls—and a somewhat mysterious ban on eating beans. Several explanations have been suggested for the bean-eating prohibition. They range from the resemblance of beans to genitals to bean eating being compared to eating a living soul. The latter interpretation regarded the wind breaking that often follows the eating of beans as proof of an extinguished breath. The book *Philosophy for Dummies* summarized the Pythagorean doctrine this way: "Everything is made of numbers, and don't eat beans because they'll do a number on you."

The oldest surviving story about Pythagoras is related to the belief in the reincarnation of the soul into other beings. This almost poetic tale comes from the sixth century BC poet Xenophanes of Colophon: "They say that once he [Pythagoras] passed by as a dog was being beaten, and pitying it spoke as follows, 'Stop, and beat it not; for the soul is that of a friend; I know it, for I heard it speak.'"

Pythagoras's unmistakable fingerprints can be found not only in the teachings of the Greek philosophers that immediately succeeded him, but all the way into the curricula of the medieval universities. The seven subjects taught in those universities were divided into the *trivium,* which included dialectic, grammar, and rhetoric, and the *quadrivium,* which included the favorite topics of the Pythagoreans—geometry, arithmetic, astronomy, and music. The celestial "harmony of the spheres"—the music supposedly performed by the planets in their orbits, which, according to his disciples, only Pythagoras could hear—has inspired poets and scientists alike. The famous astronomer Johannes Kepler (1571–1630), who discovered the laws of planetary motion, chose the title of *Harmonice Mundi* (*Harmony of the World*) for one of his most seminal works. In the Pythagorean spirit, he even developed little musical "tunes" for the different planets (as did the composer Gustav Holst three centuries later).

From the perspective of the questions that are at the focus of the present book, once we strip the Pythagorean philosophy of its mystical clothing, the skeleton that remains is still a powerful statement about mathematics, its nature, and its relation to both the physical world and the human mind. Pythagoras and the Pythagoreans were the forefathers of the search for cosmic order. They can be regarded as the founders of pure mathematics in that unlike their predecessors—the Babylonians and the Egyptians—they engaged in mathematics as an abstract field, divorced from all practical purposes. The question of whether the Pythagoreans also established mathematics as a tool for science is a trickier one. While the Pythagoreans certainly associated all phenomena with numbers, the numbers themselves—not the phenomena or their causes—became the focus of study. This was not a particularly fruitful direction for scientific research to take. Still, fundamental to the Pythagorean doctrine was the implicit belief in the existence of general, natural laws. This belief, which has become the central pillar of modern science, may have had its roots in the concept of Fate in Greek tragedy. As late as the Renaissance, this bold faith in the reality of a body of laws that can explain all phenomena was still progressing far in advance of any concrete evidence, and only Galileo, Descartes, and Newton turned it into a proposition defendable on inductive grounds.

Another major contribution attributed to the Pythagoreans was the sobering discovery that their own "numerical religion" was, in fact, pitifully unworkable. The whole numbers 1, 2, 3, . . . are insufficient even for the construction of mathematics, let alone for a description of the universe. Examine the square in figure 6, in which the length of the side is one unit, and where we denote the length of the diagonal by *d*. We can easily find the length of the diagonal, using the Pythagorean theorem in any of the two right triangles into which the square is divided. According to the theorem, the square

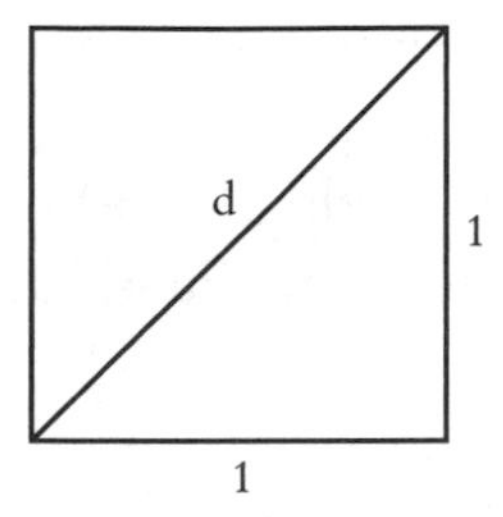

Figure 6

of the diagonal (the hypotenuse) is equal to the sum of the squares of the two shorter sides: $d^2 = 1^2 + 1^2$, or $d^2 = 2$. Once you know the square of a positive number, you find the number itself by taking the square root (e.g., if $x^2 = 9$, then the positive $x = \sqrt{9} = 3$). Therefore, $d^2 = 2$ implies $d = \sqrt{2}$ units. So the ratio of the length of the diagonal to the length of the square's side is the number $\sqrt{2}$. Here, however, came the real shock—a discovery that demolished the meticulously constructed Pythagorean discrete-number philosophy. One of the Pythagoreans (possibly Hippasus of Metapontum, who lived in the first half of the fifth century BC) managed to prove that the square root of two cannot be expressed as a ratio of any two whole numbers. In other words, even though we have an infinity of whole numbers to choose from, the search for two of them that give a ratio of $\sqrt{2}$ is doomed from the start. Numbers that can be expressed as a ratio of two whole numbers (e.g., 3/17; 2/5; 1/10; 6/1) are called *rational numbers*. The Pythagoreans proved that $\sqrt{2}$ is not a rational number. In fact, soon after the original discovery it was realized that neither are $\sqrt{3}$, $\sqrt{17}$, or the square root of any number that is not a perfect square (such as 16 or 25). The consequences were dramatic—the Pythagoreans showed that to the infinity of rational numbers we are forced to add an infinity of new kinds of numbers—ones that today we call *irrational numbers.* The importance of this discovery for the subsequent development of mathematical analysis cannot be overemphasized. Among other things, it led to the recognition of the existence of "countable" and "uncountable" infinities in the nineteenth century. The Pythagoreans, however, were so overwhelmed by this philosophical crisis that the philosopher Iamblichus reports that the man who discovered irrational numbers and disclosed their nature to "those unworthy to share in the theory" was "so hated that not only was he banned from [the Pythagoreans'] common association and way of life, but even his tomb was built, as if [their] former comrade was departed from life among mankind."

Perhaps even more important than the discovery of irrational numbers was the pioneering Pythagorean insistence on mathematical proof—a procedure based entirely on logical reasoning, by which starting from some postulates, the validity of any mathematical proposition could be unambiguously established. Prior to the Greeks, even mathematicians did not expect anyone to be interested in the least in the mental struggles that had led them to a particular discovery. If a mathematical recipe worked in practice—say for divvying up parcels of land—that was proof enough. The Greeks, on the other hand, wanted to explain why it worked. While the notion of proof may have first been introduced by the philosopher Thales of Miletus (ca. 625–547 BC), the Pythagoreans were the ones who turned this practice into an impeccable tool for ascertaining mathematical truths. The significance of this breakthrough in logic was enormous. Proofs stemming from postulates immediately put mathematics on a much firmer foundation than that of any other discipline discussed by the philosophers of the time. Once a rigorous proof, based on steps in reasoning that left no loopholes, had been presented, the validity of the associated mathematical statement was essentially unassailable. Even Arthur Conan Doyle, the creator of the world's most famous detective, recognized the special status of mathematical proof. In *A Study in Scarlet,* Sherlock Holmes declares that his conclusions are "as infallible as so many propositions of Euclid."

On the question of whether mathematics was discovered or invented, Pythagoras and the Pythagoreans had no doubt—mathematics was real, immutable, omnipresent, and more sublime than anything that could conceivably emerge from the feeble human mind. The Pythagoreans literally embedded the universe into mathematics. In fact, to the Pythagoreans, God was not a mathematician—*mathematics was God!*

The importance of the Pythagorean philosophy lies not only in its actual, intrinsic value. By setting the stage, and to some extent the agenda, for the next generation of philosophers—Plato in particular—the Pythagoreans established a commanding position in Western thought.

Into Plato's Cave

The famous British mathematician and philosopher Alfred North Whitehead (1861–1947) remarked once that "the safest generalization that can be made about the history of western philosophy is that it is all a series of footnotes to Plato."

Indeed, Plato (ca. 428–347 BC) was the first to have brought together topics ranging from mathematics, science, and language to religion, ethics, and art and to have treated them in a unified manner that essentially defined philosophy as a discipline. To Plato, philosophy was not some abstract subject, divorced from everyday activities, but rather the chief guide to how humans should live their lives, recognize truths, and conduct their politics. In particular, he maintained that philosophy can gain us access into a realm of truths that lies far beyond what we can either perceive directly with our senses or even deduce by simple common sense. Who was this relentless seeker of pure knowledge, absolute good, and eternal truths?

Plato, the son of Ariston and Perictione, was born in Athens or Aegina. Figure 7 shows a Roman herm of Plato that was most likely copied from an older, fourth century BC Greek original. His family had a long line of distinction on both sides, including such figures as Solon, the celebrated lawmaker, and Codrus, the last king of Athens. Plato's uncle Charmides and his mother's cousin Critias were old friends of the famous philosopher Socrates (ca. 470–399 BC)—a relation that in many ways defined the formative influence to which the young Plato's mind was exposed. Originally, Plato intended to enter into politics, but a series of violent actions by the political faction that courted him at the time convinced him otherwise. Later in life, this initial repulsion by politics may have encouraged Plato to outline what he regarded as the essential education for future guardians of the state. In one case, he even attempted (unsuccessfully) to tutor the ruler of Syracuse, Dionysius II.

Figure 7

Following the execution of Socrates in 399 BC, Plato embarked on extensive travel that ended only when he founded his renowned school of philosophy and science—the Academy—around 387 BC. Plato was the director (or *scholarch*) of the Academy until his death, and his nephew Speusippus succeeded him in that position. Unlike academic institutions today, the Academy was a rather informal gathering of intellectuals who, under Plato's guidance, pursued a wide variety of interests. There were no tuition fees, no prescribed curricula, and not even real faculty members. Still, there was apparently one rather unusual "entrance requirement." According to an oration by the fourth century (AD) emperor Julian the Apostate, a burdensome inscription hung over the door to Plato's Academy. While the text of the inscription does not appear in the oration, it can be found in another fourth century marginal note. The inscription read: "Let no one destitute of geometry enter." Since no fewer than eight centuries separate the establishment of the Academy and the first description of the inscription, we cannot be absolutely certain that such an inscription indeed existed. There is no doubt, however, that the sentiment expressed by this demanding requirement reflected Plato's personal opinion. In one of his famous dialogues, *Gorgias*, Plato writes: "Geometric equality is of great importance among gods and men."

The "students" in the Academy were generally self-supporting, and some of them—the great Aristotle for one—stayed there for as long as twenty years. Plato considered this long-term contact of creative minds to be the best vehicle for the production of new ideas, in topics ranging from abstract metaphysics and mathematics to ethics and politics. The purity and almost divine attributes of Plato's disciples were captured beautifully in a painting entitled *The School of Plato* by the Belgian symbolist painter Jean Delville (1867–1953). To emphasize the spiritual qualities of the students, Delville painted them in the nude, and they appear to be androgynous, because that was supposed to be the state of primordial humans.

I was disappointed to discover that archaeologists were never able to find the remains of Plato's Academy. On a trip to Greece in the summer of 2007, I looked for the next best thing. Plato mentions the Stoa of Zeus (a covered walkway built in the fifth century BC) as a favorite place to talk to friends. I found the ruins of this stoa in the northwest part of the ancient agora in Athens (which was the civic center in Plato's time; figure 8). I must say that even though the temperature reached 115 °F that day, I felt something like a shiver as I

Figure 8

walked along the same path that must have been traversed hundreds, if not thousands of times by the great man.

The legendary inscription above the Academy's door speaks loudly about Plato's attitude toward mathematics. In fact, most of the significant mathematical research of the fourth century BC was carried out by people associated in one way or another with the Academy. Yet Plato himself was not a mathematician of great technical dexterity, and his direct contributions to mathematical knowledge were probably minimal. Rather, he was an enthusiastic spectator, a motivating source of challenge, an intelligent critic, and an inspiring guide. The first century philosopher and historian Philodemus paints a clear picture: "At that time great progress was seen in mathematics, with Plato serving as the general architect setting out problems, and the mathematicians investigating them earnestly." To which the Neoplatonic philosopher and mathematician Proclus adds: "Plato . . . greatly advanced mathematics in general and geometry in particular because of his zeal for these studies. It is well known that his writings are thickly sprinkled with mathematical terms and that he everywhere tries to arouse admiration for mathematics among students of philosophy." In other words, Plato, whose mathematical knowledge was broadly up to date, could converse with the mathematicians as an equal and as a problem presenter, even though his personal mathematical achievements were not significant.

Another striking demonstration of Plato's appreciation of mathematics comes in what is perhaps his most accomplished book, *The Republic,* a mind-boggling fusion of aesthetics, ethics, metaphysics, and politics. There, in book VII, Plato (through the central figure of Socrates) outlined an ambitious plan of education designed to create utopian state rulers. This rigorous if idealized curriculum envisaged an early training in childhood imparted through play, travel, and gymnastics. After the selection of those who showed promise, the program continued with no fewer than ten years of mathematics, five years of dialectic, and fifteen years of practical experience, which included holding commands in time of war and other offices "suitable to youth." Plato gave clear explanations as to why he thought that this was the necessary training for the would-be politicians:

> What we require is that those who take office should not be lovers of rule. Otherwise there will be a contest with rival lovers. What others, then, will you compel to undertake the guardianship of the city than those who have most intelligence of the principles that are the means of good government and who possess distinctions of another kind and a life that is preferable to political life?

Refreshing, isn't it? In fact, while such a demanding program was probably impractical even in Plato's time, George Washington agreed that an education in mathematics and philosophy was not a bad idea for the politicians-to-be:

> The science of figures, to a certain degree, is not only indispensably requisite in every walk of civilized life; but investigation of mathematical truths accustoms the mind to method and correctness in reasoning, and is an employment peculiarly worthy of rational being. In a clouded state of existence, where so many things appear precarious to the bewildered research, it is here that the rational faculties find foundation to rest upon. From the high ground of mathematical and philosophical demonstration, we are insensibly led to far nobler speculations and sublimer meditations.

For the question of the nature of mathematics, even more important than Plato the mathematician or the math stimulator was Plato the philosopher of mathematics. There his trail-blazing ideas put him not only above all the mathematicians and philosophers of his generation, but identified him as an influential figure for the following millennia.

Plato's vision of what mathematics truly is makes strong reference to his famous Allegory of the Cave. There he emphasizes the doubtful validity of the information provided through the human senses. What we perceive as the real world, Plato says, is no more real than shadows projected onto the walls of a cavern. Here is the remarkable passage from *The Republic:*

> See human beings as though they were in an underground cave-like dwelling with an entrance, a long one, open to the light across the whole width of the cave. They are in it from childhood with their legs and necks in bonds so that they are fixed, seeing only in front of them, unable because of the bond to turn their heads all the way around. Their light is from a fire burning far above and behind them. Between the fire and the prisoners there is a road above, along which we see a wall, built like the partitions puppet-handlers set in front of the human beings and over which they show the puppets . . . Then also see along this wall human beings carrying all sorts of artifacts, which project above the wall, and statues of men and other animals wrought from stone, wood, and every kind of material . . . do you suppose such men would have seen anything of themselves and one another, other than the shadows cast by the fire on the side of the cave facing them?

According to Plato, we, humans in general, are no different from those prisoners in the cave who mistake the shadows for reality. (Figure 9 shows an engraving by Jan Saenredam from 1604 illustrating the allegory.) In particular, Plato stresses, mathematical truths refer not to circles, triangles, and squares that can be drawn on a piece of papyrus, or marked with a stick in the sand, but to abstract objects that dwell in an ideal world that is the home of true forms and perfections. This Platonic world of mathematical forms is distinct from the physical world, and it is in this first world that mathematical propositions, such as the Pythagorean theorem, hold true. The right triangle we might draw on paper is but an imperfect copy—an approximation—of the true, abstract triangle.

Another fundamental issue that Plato examined in some detail concerned the nature of mathematical proof as a process that is based on *postulates* and *axioms*. Axioms are basic assertions whose validity is assumed to be self-evident. For instance, the first axiom in Euclidean geometry is "Between any two points a straight line may be drawn." In *The Republic,* Plato beautifully combines the concept of postulates with his notion of the world of mathematical forms:

Figure 9

I think you know that those who occupy themselves with geometries and calculations and the like, take for granted the odd and the even [numbers], figures, three kinds of angles, and other things cognate to these in each subject; assuming these things as known, they take them as hypotheses and thenceforward they do not feel called upon to give any explanation with regard to them either to themselves or anyone else, but treat them as manifest to every one; basing themselves on these hypotheses, they proceed at once to go through the rest of the argument till they arrive, with general assent, at the particular conclusion to which their inquiry was directed. Further you know that they make use of visible figures and argue about them, but in doing so they are not thinking about these figures but of the things which they represent; thus it is the absolute square and the absolute diameter which is the object of their argument, not the diameter which they draw . . . the object of the inquirer being to see their absolute counterparts which *cannot be seen otherwise than by thought* [emphasis added].

Plato's views formed the basis for what has become known in philosophy in general, and in discussions of the nature of mathematics in particular, as *Platonism*. Platonism in its broadest sense espouses a belief in some abstract eternal and immutable realities that are entirely independent of the transient world perceived by our senses. According to Platonism, the real existence of mathematical objects is as much an objective fact as is the existence of the universe itself. Not only do the natural numbers, circles, and squares exist, but so do imaginary numbers, functions, fractals, non-Euclidean geometries, and infinite sets, as well as a variety of theorems about these entities. In short, every mathematical concept or "objectively true" statement (to be defined later) ever formulated or imagined, and an infinity of concepts and statements not yet discovered, are absolute entities, or *universals,* that can neither be created nor destroyed. They exist independently of our knowledge of them. Needless to say, these objects are not physical—they live in an autonomous world of timeless essences. Platonism views mathematicians as explorers of foreign lands; they can only discover mathematical truths, not invent them. In the same way that America was already there long before Columbus (or Leif Ericson) discovered it, mathematical theorems existed in the Platonic world before the Babylonians ever initiated mathematical studies. To Plato, the only things that truly and wholly exist are those abstract forms and ideas of mathematics, since only in mathematics, he maintained, could we gain absolutely certain and objective knowledge. Consequently, in Plato's mind, mathematics becomes closely associated with the divine. In the dialogue *Timaeus,* the creator god uses mathematics to fashion the world, and in *The Republic,* knowledge of mathematics is taken to be a crucial step on the pathway to knowing the divine forms. Plato does not use mathematics for the formulation of some laws of nature that are testable by experiments. Rather, for him, the mathematical character of the world is simply a consequence of the fact that "God always geometrizes."

Plato extended his ideas on "true forms" to other disciplines as well, in particular to astronomy. He argued that in true astronomy "we must leave the heavens alone" and not attempt to account for the arrangements and the apparent motions of the visible stars. Instead, Plato regarded true astronomy as a science dealing with the laws of motion in some ideal, mathematical world, for which the observable heaven is a mere illustration (in the same way that geometrical figures drawn on papyrus only illustrate the true figures).

Plato's suggestions for astronomical research are considered controversial even by some of the most devout Platonists. Defenders of his ideas argue that what Plato really means is not that true astronomy should concern itself with some ideal heaven that has nothing to do with the observable one, but that it should deal with the real motions of celestial bodies as opposed to the apparent motions as seen from Earth. Others point out, however, that too literal an adoption of Plato's dictum would have seriously impeded the development of observational astronomy as a science. Be the interpretation of Plato's attitude toward astronomy as it may, Platonism has become one of the leading dogmas when it comes to the foundations of mathematics.

But does the Platonic world of mathematics really exist? And if it does, where exactly is it? And what are these "objectively true" statements that inhabit this world? Or are the mathematicians who adhere to Platonism simply expressing the same type of romantic belief that has been attributed to the great Renaissance artist Michelangelo? According to legend, Michelangelo believed that his magnificent sculptures already existed inside the blocks of marble and that his role was merely to uncover them.

Modern-day Platonists (yes, they definitely exist, and their views will be described in more detail in later chapters) insist that the Platonic world of mathematical forms is real, and they offer what they regard as concrete examples of objectively true mathematical statements that reside in this world.

Take the following easy-to-understand proposition: Every even integer greater than 2 can be written as the sum of two primes (numbers divisible only by one and themselves). This simple-sounding statement is known as the Goldbach conjecture, since an equivalent conjecture appeared in a letter written by the Prussian amateur mathematician Christian Goldbach (1690–1764) on June 7, 1742. You can easily verify the validity of the conjecture for the first few even numbers: $4 = 2 + 2$; $6 = 3 + 3$; $8 = 3 + 5$; $10 = 3 + 7$ (or $5 + 5$); $12 = 5 + 7$; $14 = 3 + 11$ (or $7 + 7$); $16 = 5 + 11$ (or $3 + 13$); and so on. The statement is so simple that the British mathematician G. H. Hardy declared that "any fool could have guessed it." In fact, the great French mathematician and philosopher René Descartes had anticipated this conjecture before Goldbach. Proving the conjecture, however, turned out to be quite a different matter. In 1966 the Chinese mathematician Chen Jingrun made a significant step toward a proof. He managed to show that every sufficiently large even integer is the sum of two numbers, one of which is a prime and the other has at most two prime factors. By the end of 2005, the Portuguese researcher Tomás Oliveira e Silva had shown the conjecture to be true for numbers up to 3×10^{17} (three hundred thousand trillion). Yet, in spite of enormous efforts by many talented mathematicians, a general proof remains elusive at the time of this writing. Even the additional temptation of a $1 million prize offered between March 20, 2000, and March 20, 2002 (to help publicize a novel entitled *Uncle Petros and Goldbach's Conjecture*), did not produce the desired result. Here, however, comes the crux of the meaning of "objective truth" in mathematics. Suppose that a rigorous proof will actually be formulated in 2016. Would we then be able to say that the statement was already true when Descartes first thought about it? Most people would agree that this question is silly. Clearly, if the proposition is proven to be true, then it has *always* been true, even before we knew it to be true. Or, let's look at another innocent-looking example known as *Catalan's conjecture*. The numbers 8 and 9 are consecutive whole numbers, and each of them is equal to a pure power, that is $8 = 2^3$ and $9 = 3^2$. In 1844, the Belgian mathematician Eugène Charles Catalan (1814–94) conjectured that among all the possible powers of whole numbers, the only pair of consecutive numbers (excluding 0 and 1) is 8 and 9. In other words, you can spend your life writing down all the pure powers that exist. Other than 8 and 9, you will find no other two numbers that differ by only 1. In 1342, the Jewish-French philosopher and mathematician Levi Ben Gerson (1288–1344) actually proved a small part of the conjecture—that 8 and 9 are the only powers of 2 and 3 differing by 1. A major step forward was taken by the mathematician Robert Tijdeman in 1976. Still, the proof of the general form of Catalan's conjecture stymied the best mathematical minds for more than 150 years. Finally, on April 18, 2002, the Romanian mathematician Preda Mihailescu presented a complete proof of the conjecture. His proof was published in 2004 and is now fully accepted. Again you may ask: When did Catalan's conjecture become true? In 1342? In 1844? In 1976? In 2002? In 2004? Isn't it obvious that the statement was always true, only that we didn't know it to be true? These are the types of truths Platonists would refer to as "objective truths."

Some mathematicians, philosophers, cognitive scientists, and other "consumers" of mathematics (e.g., computer scientists) regard the Platonic world as a figment of the imagination of too-dreamy minds (I shall describe this perspective and other dogmas in detail later in the book). In fact, in 1940, the famous historian of mathematics Eric Temple Bell (1883–1960) made the following prediction:

> According to the prophets, the last adherent of the Platonic ideal in mathematics will have joined the dinosaurs by the year 2000. Divested of its mythical raiment of eternalism, mathematics will then be recognized for what it has always been, a humanly constructed language devised by human beings for definite ends prescribed by themselves. The last temple of an absolute truth will have vanished with the nothing it enshrined.

Bell's prophecy proved to be wrong. While dogmas that are diametrically opposed (but in different directions) to Platonism have emerged, those have not fully won the minds (and hearts!) of all mathematicians and philosophers, who remain today as divided as ever.

Suppose, however, that Platonism had won the day, and we had all become wholehearted Platonists. Does Platonism actually explain the "unreasonable effectiveness" of mathematics in describing our world? Not really. Why should physical reality behave according to laws that reside in the abstract Platonic world? This was, after all, one of Penrose's mysteries, and Penrose is a devout Platonist himself. So for the moment we have to accept the fact that even if we were to embrace Platonism, the puzzle of the powers of mathematics would remain unsolved. In Wigner's words: "It is difficult to avoid the impression that a miracle confronts us here, comparable in its striking nature to the miracle that the human mind can string a thousand arguments together without getting itself into contradictions."

To fully appreciate the magnitude of this miracle, we have to delve into the lives and legacies of some of the miracle workers themselves—the minds behind the discoveries of a few of those incredibly precise mathematical laws of nature.

CHAPTER 3

MAGICIANS: THE MASTER AND THE HERETIC

Unlike the Ten Commandments, science was not handed to humankind on imposing tablets of stone. The history of science is the story of the rise and fall of numerous speculations, hypotheses, and models. Many seemingly clever ideas turned out to be false starts or led down blind alleys. Some theories that were taken to be ironclad at the time later dissolved when put to the fiery test of subsequent experiments and observations, only to become entirely obsolete. Even the extraordinary brainpower of the originators of some conceptions did not make those conceptions immune to being superseded. The great Aristotle, for instance, thought that stones, apples, and other heavy objects fall down because they seek their natural place, which is at the center of Earth. As they approached the ground, Aristotle argued, these bodies increased their speed because they were happy to return home. Air (and fire), on the other hand, moved upward because the air's natural place was with the heavenly spheres. All objects could be assigned a nature based on their perceived relation to the most basic constituents—earth, fire, air, and water. In Aristotle's words:

> Some existing things are natural, while others are due to other causes. Those that are natural are . . . the simple bodies such as earth, fire, air and water . . . all these things evidently differ from those that are not naturally constituted, since each

> of them has within itself a principle of motion and stability in place . . . A nature is a type of principle and cause of motion and stability within these things to which it primarily belongs . . . The things that are in accordance with nature include both these and whatever belongs to them in their own right, as traveling upward belongs to fire.

Aristotle even made an attempt to formulate a quantitative law of motion. He asserted that heavier objects fall faster, with the speed being directly proportional to the weight (that is, an object two times heavier than another was supposed to fall at twice the speed). While everyday experience might have made this law seem reasonable enough—a brick was indeed observed to hit the ground earlier than a feather dropped from the same height—Aristotle never examined his quantitative statement more precisely. Somehow, it either never occurred to him, or he did not consider it necessary, to check whether two bricks tied together indeed fall twice as fast as a single brick. Galileo Galilei (1564–1642), who was much more mathematically and experimentally oriented, and who showed little respect for the happiness of falling bricks and apples, was the first to point out that Aristotle got it completely wrong. Using a clever thought experiment, Galileo was able to demonstrate that Aristotle's law just didn't make any sense, because it was logically inconsistent. He argued as follows: Suppose you tie together two objects, one heavier than the other. How fast would the combined object fall compared to each of its two constituents? On one hand, according to Aristotle's law, you might conclude that it would fall at some intermediate speed, because the lighter object would slow down the heavier one. On the other, given that the combined object is actually heavier than its components, it should fall even faster than the heavier of the two, leading to a clear contradiction. The only reason that a feather falls on Earth more gently than a brick is that the feather experiences greater air resistance—if dropped from the same height in a vacuum, they would hit the ground simultaneously. This fact has been demonstrated in numerous experiments, none more dramatic than the one performed by Apollo 15 astronaut David Randolph Scott. Scott—the seventh person to walk on the Moon—simultaneously dropped a hammer from one hand and a feather from the other. Since the Moon lacks a substantial atmosphere, the hammer and the feather struck the lunar surface at the same time.

The amazing fact about Aristotle's false law of motion is not that it was wrong, but that it was accepted for almost two thousand years. How could a flawed idea enjoy such a remarkable longevity? This was a case of a "perfect storm"—three different forces combining to create an unassailable doctrine. First, there was the simple fact that in the absence of precise measurements, Aristotle's law seemed to agree with experience-based common sense—sheets of papyrus did hover about, while lumps of lead did not. It took Galileo's genius to argue that common sense could be misleading. Second, there was the colossal weight of Aristotle's almost unmatched reputation and authority as a scholar. After all, this was the man who laid out the foundations for much of Western intellectual culture. Whether it was the investigation of all natural phenomena or the bedrock of ethics, metaphysics, politics, or art, Aristotle literally wrote the book. And that was not all. Aristotle in some sense even taught us *how* to think, by introducing the first formal studies of logic. Today, almost every child at school recognizes Aristotle's pioneering, virtually complete system of logical inference, known as a *syllogism:*

1. Every Greek is a person.
2. Every person is mortal.
3. Therefore every Greek is mortal.

The third reason for the incredible durability of Aristotle's incorrect theory was the fact that the Christian church adopted this theory as a part of its own official orthodoxy. This acted as a deterrent against most attempts to question Aristotle's assertions.

In spite of his impressive contributions to the systemization of deductive logic, Aristotle is not noted for his mathematics. Somewhat surprisingly perhaps, the man who essentially established science as an organized enterprise did not care as much (and certainly not as much as Plato) for mathematics and was rather weak in physics. Even though Aristotle recognized the importance of numerical and geometrical relationships in the sciences, he still regarded mathematics as an abstract discipline, divorced from physical reality. Consequently, while there is no doubt that he was an intellectual powerhouse, Aristotle does *not* make my list of mathematical "magicians."

I am using the term "magicians" here for those individuals who could pull rabbits out of literally empty hats; those who discovered never-before-thought-of connections between mathematics and nature; those who were able to observe complex natural phenomena and to distill from them crystal-clear mathematical laws. In some cases, these superior thinkers even used their experiments and observations to advance their mathematics. The question of the unreasonable effectiveness of mathematics in explaining nature would never have arisen were it not for these magicians. This enigma was born directly out of the miraculous insights of these researchers.

No single book can do justice to all the superb scientists and mathematicians who have contributed to our understanding of the universe. In this chapter and the following one I intend to concentrate on only four of those giants of past centuries, about whose status as magicians there can be no doubt—some of the crème de la crème of the scientific world. The first magician on my list is best remembered for a rather unusual event—for dashing stark naked through the streets of his hometown.

Give Me a Place to Stand and I Will Move the Earth

When the historian of mathematics Eric Temple Bell had to decide whom to name as his top three mathematicians, he concluded:

> Any list of the three "greatest" mathematicians of all history would include the name of Archimedes. The other two usually associated with him are Newton (1642–1727) and Gauss (1777–1855). Some, considering the relative wealth—or poverty—of mathematics and physical science in the respective ages in which these giants lived, and estimating their achieve-

ments against the background of their times, would put Archimedes first.

Archimedes (287–212 BC; figure 10 shows a bust claimed to represent Archimedes, but which may in fact be that of a Spartan king) was indeed the Newton or Gauss of his day; a man of such brilliance, imagination, and insight that both his contemporaries and the generations that followed him uttered his name in awe and reverence. Even though he is better known for his ingenious inventions in engineering, Archimedes was primarily a mathematician, and in his mathematics he was centuries ahead of his time. Unfortunately, little is known about Archimedes' early life or his family. His first biography, written by one Heracleides, has not survived, and the few details that we do know about his life and violent death come primarily from the writings of the Roman historian Plutarch. Plutarch (ca. AD 46–120) was, in fact, more interested in the military accomplishments of the Roman general Marcellus, who conquered Archimedes' home town of Syracuse in 212 BC. Fortunately for the history of mathematics, Archimedes had given Marcellus such a tremendous headache during the siege of Syracuse that the three major historians of the period, Plutarch, Polybius, and Livy, couldn't ignore him.

Figure 10

Archimedes was born in Syracuse, then a Greek settlement in Sicily. According to his own testimony, he was the son of the astronomer Phidias, about whom little is known beyond the fact that he had estimated the ratio of the diameters of the Sun and the Moon. Archimedes may have also been related in some way to King Hieron II, himself the illegitimate son of a nobleman (by one of the latter's female slaves). Irrespective of whichever ties Archimedes might have had with the royal family, both the king and his son, Gelon, always held Archimedes in high regard. As a youth, Archimedes spent some time in Alexandria, where he studied mathematics, before returning to a life of extensive research in Syracuse.

Archimedes was truly a mathematician's mathematician. According to Plutarch, he regarded as sordid and ignoble "every art directed to use and profit, and he only strove after those things which, in their beauty and excellence, remain beyond all contact with the common needs of life." Archimedes' preoccupation with abstract mathematics and the level to which he was consumed by it apparently went much farther even than the enthusiasm commonly exhibited by practitioners of this discipline. Again according to Plutarch:

> Continually bewitched by a Siren who always accompanied him, he forgot to nourish himself and omitted to care for his body; and when, as would often happen, he was urged by force to bathe and anoint himself, he would still be drawing geometrical figures in the ashes or with his fingers would draw lines on his anointed body, being possessed by a great ecstasy and in truth a thrall to the Muses.

In spite of his contempt for applied mathematics, and the little importance that Archimedes himself attached to his engineering ideas, his resourceful inventions gained him even more popular fame than his mathematical genius.

The best-known legend about Archimedes further enhances his image as the stereotypical absentminded mathematician. This amusing story was first told by the Roman architect Vitruvius in the first century BC, and it goes like this: King Hieron wanted to consecrate a gold wreath to the immortal gods. When the wreath was delivered to the king, it was equal in weight to the gold furnished for its creation. The king was nonetheless suspicious that a certain amount of gold had been replaced by silver of the same weight. Not being able to substantiate his distrust, the king turned for advice to the master of mathematicians—Archimedes. One day, the legend continued, Archimedes stepped into a bath, while still engrossed in the problem of how to uncover potential fraud with the wreath. As he immersed himself in the water, however, he realized that his body displaced a certain volume of water, which overflowed the tub's edge. This immediately triggered a solution in his head. Overwhelmed with joy, Archimedes jumped out of the tub and ran naked in the street shouting "*Eureka, eureka!*" ("I have found it, I have found it!").

Another famous Archimedean exclamation, "Give me a place to stand and I will move the Earth," is currently featured (in one version or another) on more than 150,000 Web pages found in a Google search. This bold proclamation, sounding almost like the vision statement of a large corporation, has been cited by Thomas Jefferson, Mark Twain, and John F. Kennedy and it was even featured in a poem by Lord Byron. The phrase was apparently the culmination of Archimedes' investigations into the problem of moving a given weight with a given force. Plutarch tells us that when King Hieron asked for a practical demonstration of Archimedes' ability to manipulate a large weight with a small force, Archimedes managed—using a compound pulley—to launch a fully loaded ship into the sea. Plutarch adds in admiration that "he drew the ship along smoothly and safely as if she were moving through the sea." Slightly modified versions of the same legend appear in other sources. While it is difficult to believe that Archimedes could have actually moved an entire ship with the mechanical devices available to him at the time, the legends leave little room for doubt that he gave some impressive demonstration of an invention that enabled him to maneuver heavy weights.

Archimedes made many other peacetime inventions, such as a hydraulic screw for raising water and a planetarium that demonstrated the motions of the heavenly bodies, but he became most famous in antiquity for his role in the defense of Syracuse against the Romans.

Wars have always been popular with historians. Consequently, the events of the Roman siege on Syracuse during the years 214–212 BC have been lavishly chronicled by many historians. The Roman general Marcus Claudius Marcellus (ca. 268–208 BC), by then of considerable military fame, anticipated a rapid victory. He apparently failed to consider a stubborn King Hieron, aided by a mathematical and engineering genius. Plutarch gives a vivid description of the havoc that Archimedes' machines inflicted upon the Roman forces:

> He [Archimedes] at once shot against the land forces all sorts of missile weapons, and immense masses of stone that came down with incredible noise and violence; against which no man could stand; for they knocked down those upon whom they fell in heaps, breaking all their ranks and files. At the same time huge poles thrust out from the walls over the ships sunk some by great weights which they let down from on high upon them; others they lifted up into the air by an iron hand or beak like a crane's beak and, when they had drawn them up by the prow, and set them on end upon the poop, they plunged them to the bottom of the sea . . . A ship was frequently lifted up to a great height in the air (a dreadful thing to behold), and was rolled to and fro, and kept swinging, until the mariners were all thrown out, when at length it was dashed against the rocks, or let fall.

The fear of the Archimedean devices became so extreme that "if they [the Roman soldiers] did but see a piece of rope or wood projecting above the wall, they would cry 'there it is again,' declaring that Archimedes was setting some engine in motion against them, and would turn their backs and run away." Even Marcellus was deeply impressed, complaining to his own crew of military engineers: "Shall we not make an end of fighting against this geometrical Briareus [the hundred-armed giant, son of Uranus and Gaia] who, sitting at ease by the sea, plays pitch and toss with our ships to our confusion, and by the multitude of missiles that he hurls at us outdoes the hundred-handed giants of mythology?"

According to another popular legend that appeared first in the writings of the great Greek physician Galen (ca. AD 129–200), Archimedes used an assembly of mirrors that focused the Sun's rays to burn the Roman ships. The sixth century Byzantine architect Anthemius of Tralles and a number of twelfth century historians repeated this fantastic story, even though the actual feasibility of such a feat remains uncertain. Still, the collection of almost mythological tales does provide us with rich testimony as to the veneration that "the wise one" inspired in later generations.

As I noted earlier, Archimedes himself—that highly esteemed "geometrical Briareus"—attached no particular significance to all of his military toys; he basically regarded them as diversions of geometry at play. Unfortunately, this aloof attitude may have eventually cost Archimedes his life. When the Romans finally captured Syracuse, Archimedes was so busy drawing his geometrical diagrams on a dust-filled tray that he failed to notice the tumult of war. According to some accounts, when a Roman soldier ordered Archimedes to follow him to Marcellus, the old geometer retorted indignantly: "Fellow, stand away from my diagram." This reply infuriated the soldier to such a degree that, disobeying his commander's specific orders, he unsheathed his sword and slew the greatest mathematician of antiquity. Figure 11 shows what is believed to be a reproduction (from the eighteenth century) of a mosaic found in Herculaneum depicting the final moments in the life of "the master."

Archimedes' death marked, in some sense, the end of an extraordinarily vibrant era in the history of mathematics. As the British mathematician and philosopher Alfred North Whitehead remarked:

> The death of Archimedes at the hands of a Roman soldier is symbolical of a world change of the first magnitude. The Romans were a great race, but they were cursed by the sterility which waits upon practicality. They were not dreamers enough to arrive at new points of view, which could give more fundamental control over the forces of nature. No Roman lost his life because he was absorbed in the contemplation of a mathematical diagram.

Figure 11

Fortunately, while details of Archimedes' life are scarce, many (but not all) of his incredible writings have survived. Archimedes had a habit of sending notes on his mathematical discoveries to a few mathematician friends or to people he respected. The exclusive list of correspondents included (among others) the astronomer Conon of Samos, the mathematician Eratosthenes of Cyrene, and the king's son, Gelon. After Conon's death, Archimedes sent a few notes to Conon's student, Dositheus of Pelusium.

Archimedes' opus covers an astonishing range of mathematics and physics. Among his many achievements: He presented general methods for finding the areas of a variety of plane figures and the volumes of spaces bounded by all kinds of curved surfaces. These included the areas of the circle, segments of a parabola and of a spiral, and volumes of segments of cylinders, cones, and other figures generated by the revolution of parabolas, ellipses, and hyperbolas. He showed that the value of the number π, the ratio of the circumference of a circle to its diameter, has to be larger than $3\frac{10}{71}$ and smaller than $3\frac{1}{7}$. At a time when no method existed to describe very large numbers, he invented a system that allowed him not only to write down, but also to manipulate numbers of any magnitude. In physics, Archimedes discovered the laws governing floating bodies, thus establishing the science of hydrostatics. In addition, he calculated the centers of gravity of many solids and formulated the mechanical laws of levers. In astronomy, he performed observations to determine the length of the year and the distances to the planets.

The works of many of the Greek mathematicians were characterized by originality and attention to detail. Still, Archimedes' methods of reasoning and solution truly set him apart from all of the scientists of his day. Let me describe here only three representative examples that give the flavor of Archimedes' inventiveness. One appears at first blush to be nothing more than an amusing curiosity, but a closer examination reveals the depth of his inquisitive mind. The other two illustrations of the Archimedean methods demonstrate such ahead-of-his-time thinking that they immediately elevate Archimedes to what I dub the "magician" status.

Archimedes was apparently fascinated by big numbers. But very large numbers are clumsy to express when written in ordinary notation (try writing a personal check for $8.4 trillion, the U.S. national debt in July 2006, in the space allocated for the figure amount). So Archimedes developed a system that allowed him to represent numbers with 80,000 trillion digits. He then used this system in an original treatise entitled *The Sand Reckoner,* to show that the total number of sand grains in the world was not infinite.

Even the introduction to this treatise is so illuminating that I will reproduce a part of it here (the entire piece was addressed to Gelon, the son of King Hieron II):

> There are some, king Gelon, who think that the number of the sand is infinite in multitude; and I mean by the sand not only that which exists about Syracuse and the rest of Sicily but also that which is found in every region whether inhabited or uninhabited. Again there are some who, without regarding it as infinite, yet think that no number has been named which is great enough to exceed its multitude. And it is clear that they who hold this view, if they imagined a mass made up of sand in other respects as large as the mass of the earth, including in it all the

> seas and the hollows of the earth filled up to a height equal to that of the highest of the mountains, would be many times further still from recognizing that any number could be expressed which exceeds the multitude of the sand so taken. But I will try to show you by means of geometrical proofs, which you will be able to follow, that, of the numbers named by me and given in the work which I sent to Zeuxippus [a work that has unfortunately been lost], some exceed not only the number of the mass of sand equal in magnitude to the earth filled up in the way described, but also that of a mass equal in magnitude to the universe. Now you are aware that "universe" is the name given by most astronomers to the sphere whose center is the center of the earth and whose radius is equal to the straight line between the center of the Sun and the center of the Earth. This is the common account, as you have heard from astronomers. But Aristarchus of Samos brought out a book consisting of some hypotheses, in which the premises lead to the result that the universe is many times greater than that now so called. His hypotheses are that the fixed stars and the Sun remain unmoved, that the Earth revolves about the Sun in the circumference of a circle, the Sun lying in the middle of the orbit.

This introduction immediately highlights two important points: (1) Archimedes was prepared to question even very popular beliefs (such as that there is an infinity of grains of sand), and (2) he treated with respect the heliocentric theory of the astronomer Aristarchus (later in the treatise he actually corrected one of Aristarchus's hypotheses). In Aristarchus's universe the Earth and the planets revolved around a stationary Sun that was located at the center (remember that this model was proposed 1,800 years before Copernicus!). After these preliminary remarks, Archimedes starts to address the problem of the grains of sand, progressing by a series of logical steps. First he estimates how many grains placed side by side it would take to cover the diameter of a poppy seed. Then, how many poppy seeds would fit in the breadth of a finger; how many fingers in a stadium (about 600 feet);

and continuing up to ten billion stadia. Along the way, Archimedes invents a system of indices and a notation that, when combined, allow him to classify his gargantuan numbers. Since Archimedes assumed that the sphere of the fixed stars is less than ten million times larger than the sphere containing the orbit of the Sun (as seen from Earth), he found the number of grains in a sand-packed universe to be less than 10^{63} (one followed by sixty-three zeros). He then concluded the treatise with a respectful note to Gelon:

> I conceive that these things, king Gelon, will appear incredible to the great majority of people who have not studied mathematics, but that to those who are conversant therewith and have given thought to the question of the distances and sizes of the Earth and Sun and the Moon and the whole universe the proof will carry conviction. And it was for this reason that I thought the subject would not be inappropriate for your consideration.

The beauty of *The Sand Reckoner* lies in the ease with which Archimedes hops from everyday objects (poppy seeds, sand, fingers) to abstract numbers and mathematical notation, and then back from those to the sizes of the solar system and the universe as a whole. Clearly, Archimedes possessed such intellectual flexibility that he could comfortably use his mathematics to discover unknown properties of the universe, and use the cosmic characteristics to advance arithmetical concepts.

Archimedes' second claim to the title of "magician" comes from the method that he used to arrive at many of his outstanding geometrical theorems. Very little was known about this method and about Archimedes' thought process in general until the twentieth century. His concise style gave away very few clues. Then, in 1906, a dramatic discovery opened a window into the mind of this genius. The story of this discovery reads so much like one of the historical mystery novels by the Italian author and philosopher Umberto Eco that I feel compelled to take a brief detour to tell it.

The Archimedes Palimpsest

Sometime in the tenth century, an anonymous scribe in Constantinople (today's Istanbul) copied three important works of Archimedes: *The Method, Stomachion,* and *On Floating Bodies.* This was probably part of a general interest in Greek mathematics that was largely sparked by the ninth century mathematician Leo the Geometer. In 1204, however, soldiers of the Fourth Crusade were lured by promises of financial support to sack Constantinople. In the years that followed, the passion for mathematics faded, while the schism between the Catholic Church of the west and the Orthodox Church of the east became a *fait accompli.* Sometime before 1229, the manuscript containing Archimedes' works underwent a catastrophic act of recycling—it was unbound and washed so the parchment leaves could be reused for a Christian prayer book. The scribe Ioannes Myronas finished copying the prayer book on April 14, 1229. Fortunately, the washing of the original text did not obliterate the writing completely. Figure 12 shows a page from the manuscript, with the horizontal lines representing the prayer texts and the faint vertical lines the mathematical contents. By the sixteenth century, the palimpsest—the recycled document—somehow made its way to the Holy Land, to the monastery in St. Sabas, east of Bethlehem. In the early nineteenth century, this monastery's library contained no fewer than a thousand manuscripts. Still, for reasons that are not entirely clear, the Archimedes palimpsest was moved yet again to Constantinople. Then, in the 1840s, the famous German biblical scholar Constantine Tischendorf (1815–74), the discoverer of one of the earliest Bible manuscripts, visited the Metochion of the Holy Sepulcher in Constantinople (a daughter house of the Greek Patriarchate in Jerusalem) and saw the palimpsest there. Tischendorf must have found the partially visible underlying mathematical text quite intriguing, since he apparently tore off and stole one page from the manuscript. Tischendorf's estate sold that page in 1879 to the Cambridge University Library.

In 1899, the Greek scholar A. Papadopoulos-Kerameus cataloged all the manuscripts that were housed in the Metochion, and the Archi-

Figure 12

medes manuscript appeared as Ms. 355 on his list. Papadopoulos-Kerameus was able to read a few lines of the mathematical text, and perhaps realizing their potential importance, he printed those lines in his catalog. This was a turning point in the saga of this manuscript. The mathematical text in the catalog was brought to the attention of the Danish philologist Johan Ludvig Heiberg (1854–1928). Recognizing the text as belonging to Archimedes, Heiberg traveled to Istanbul in 1906, examined and photographed the palimpsest, and a year later announced his sensational discovery—two never-before-seen treatises of Archimedes and one previously known only from its Latin translation. Even though Heiberg was able to read and later publish parts of the manuscript in his book on Archimedes' works, serious gaps remained. Unfortunately, sometime after 1908, the manuscript disappeared from Istanbul under mysterious circumstances, only to reappear in the possession of a Parisian family, who claimed to have had it since the 1920s. Improperly stored, the palimpsest had suffered some irreversible mold damage, and three pages previously transcribed by Heiberg were missing altogether. In addition, later than 1929 someone painted four Byzantine-style illuminations over four pages. Eventually, the French family that held the manuscript sent it to Christie's for auction. Ownership of the manuscript was disputed in federal court in New York in 1998. The Greek Orthodox Patriarchate of Jerusalem claimed that the manuscript had been stolen in the 1920s from one of its monasteries, but the judge ruled in favor of Christie's. The palimpsest was subsequently auctioned at Christie's on October 29, 1998, and it fetched $2 million from an anonymous buyer. The owner deposited the Archimedes manuscript at the Walters Art Museum in Baltimore, where it is still undergoing intensive conservation work and thorough examination. Modern imaging scientists have in their arsenal tools not available to the earlier researchers. Ultraviolet light, multispectral imaging, and even focused X-rays (to which the palimpsest was exposed at the Stanford Linear Accelerator Center) have already helped to decipher parts of the manuscript that had not been previously revealed. At the time of this writing, the careful scholarly study of the Archimedes manuscript is ongoing. I was fortunate enough to meet with the palimpsest's forensic team, and figure 13 shows me next to the experimental setup as it illuminates one page of the palimpsest at different wavelengths.

The drama surrounding the palimpsest is only fitting for a document that gives us an unprecedented glimpse of the great geometer's method.

The Method

When you read any book of Greek geometry, you cannot help but be impressed with the economy of style and the precision with which the theorems were stated and proved more than two millennia ago. What those books don't normally do, however, is give you clear hints as to how those theorems were conceived in the first place. Archimedes' exceptional document *The Method* partially fills in this intriguing gap—it reveals how Archimedes himself became convinced of the truth of certain theorems before he knew how to prove them. Here is part of what he wrote to the mathematician Eratosthenes of Cyrene (ca. 276–194 BC) in the introduction:

Figure 13

I will send you the proofs of these theorems in this book. Since, as I said, I know that you are diligent, an excellent teacher of philosophy, and greatly interested in any mathematical investigations that may come your way, I thought it might be appropriate to write down and set forth for you in this same book a certain special method, by means of which you will be enabled to recognize certain mathematical questions *with the aid of mechanics* [emphasis added]. I am convinced that this is no less useful for finding the proofs of these same theorems. For some things, which first became clear to me by the mechanical method, were afterwards proved geometrically, because their investigation by the said method does not furnish an actual demonstration. For it is easier to supply the proof when we have previously acquired, by the method, some knowledge of the questions than it is to find it without any previous knowledge.

Archimedes touches here on one of the most important points in scientific and mathematical research—it is often more difficult to discover what the important questions or theorems are than it is to find solutions to known questions or proofs to known theorems. So how did Archimedes discover some new theorems? Using his masterful understanding of mechanics, equilibrium, and the principles of the lever, he weighed in his mind solids or figures whose volumes or areas he was attempting to find against ones he already knew. After determining in this way the answer to the unknown area or volume, he found it much easier to prove geometrically the correctness of that answer. Consequently *The Method* starts with a number of statements on centers of gravity and only then proceeds to the geometrical propositions and their proofs.

Archimedes' method is extraordinary in two respects. First, he has essentially introduced the concept of a *thought experiment* into rigorous research. The nineteenth century physicist Hans Christian Ørsted first dubbed this tool—an imaginary experiment conducted in lieu of a real one—*Gedankenexperiment* (in German: "an experiment conducted in the thought"). In physics, where this concept has been extremely fruitful, thought experiments are used either to provide insights prior to performing actual experiments or in cases where the real experiments cannot be carried out. Second, and more important, Archimedes freed mathematics from the somewhat artificial chains that Euclid and Plato had put on it. To these two individuals, there was one way, and one way only, to do mathematics. You had to start from the axioms and proceed by an inexorable sequence of logical steps, using well-prescribed tools. The free-spirited Archimedes, on the other hand, simply utilized every type of ammunition he could think of to formulate new problems and to solve them. He did not hesitate to explore and exploit the connections between the abstract mathematical objects (the Platonic forms) and physical reality (actual solids or flat objects) to advance his mathematics.

A final illustration that further solidifies Archimedes' status as a magician is his anticipation of *integral and differential calculus* —a branch of mathematics formally developed by Newton (and independently by the German mathematician Leibniz) only at the end of the seventeenth century.

The basic idea behind the process of *integration* is quite simple (once it is pointed out!). Suppose that you need to determine the area of the segment of an ellipse. You could divide the area into many rectangles of equal width and sum up the areas of those rectangles (figure 14). Clearly, the more rectangles you use, the closer the sum will get to the actual area of the segment. In other words, the area of the segment is really equal to the limit that the sum of rectangles approaches as the number of rectangles increases to infinity. Finding this limit is called *integration.* Archimedes used some version of the method I have just described to find the volumes and surface areas of the sphere, the cone, and of ellipsoids and paraboloids (the solids you get when you revolve ellipses or parabolas about their axes).

In *differential calculus,* one of the main goals is to find the slope of a straight line that is tangent to a curve at a given point, that is, the line that touches the curve only at that point. Archimedes solved this problem for the special case of a spiral, thereby peeping into the future work of Newton and Leibniz. Today, the areas of differential and integral calculus and their daughter branches form the basis on which most mathematical models are built, be it in physics, engineering, economics, or population dynamics.

Archimedes changed the world of mathematics and its perceived relation to the cosmos in a profound way. By displaying an astounding combination of theoretical and practical interests, he provided the first empirical, rather than mythical, evidence for an apparent mathematical design of nature. The perception of mathematics being the language of the universe, and therefore the concept of God as a mathematician, was born in Archimedes' work. Still, there was something that Archimedes did not do—he never discussed the limitations of his

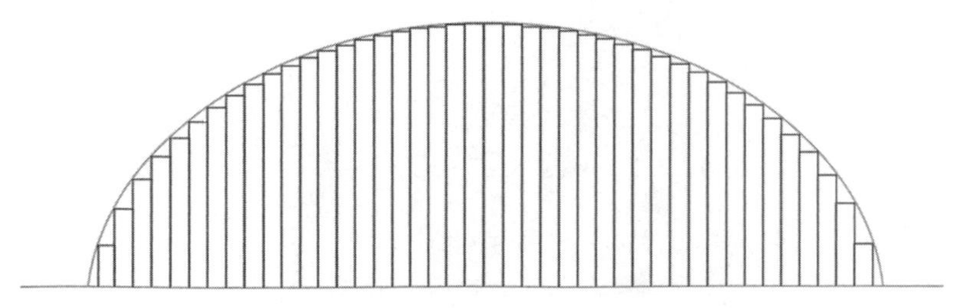

Figure 14

mathematical models when applied to actual physical circumstances. His theoretical discussions of levers, for instance, assumed that they were infinitely rigid and that rods had no weight. Consequently, he opened the door, to some extent, to the "saving the appearances" interpretation of mathematical models. This was the notion that mathematical models may only represent what is observed by humans, rather than describing the actual, true, physical reality. The Greek mathematician Geminus (ca. 10 BC–AD 60) was the first to discuss in some detail the difference between mathematical modeling and physical explanations in relation to the motion of celestial bodies. He distinguished between astronomers (or mathematicians), who, according to him, had only to suggest models that would *reproduce* the observed motions in the heavens, and physicists, who had to find *explanations* for the real motions. This particular distinction was going to come to a dramatic head at the time of Galileo, and I will return to it later in this chapter.

Somewhat surprisingly perhaps, Archimedes himself considered as one of his most cherished accomplishments the discovery that the volume of a sphere inscribed in a cylinder (figure 15) is always ⅔ of the volume of the cylinder. He was so pleased with this result that he requested it be engraved on his tombstone. Some 137 years after Archimedes' death, the famous Roman orator Marcus Tullius Cicero (ca. 106–43 BC) discovered the great mathematician's grave. Here is Cicero's rather moving description of the event:

Figure 15

> When I was a quaestor in Sicily I managed to track down his [Archimedes'] grave. The Syracusans knew nothing about it, and indeed denied that any such thing existed. But there it was, completely surrounded and hidden by bushes of brambles and thorns. I remembered having heard of some simple lines of verse which had been inscribed on his tomb, referring to a sphere and a cylinder modeled in stone on top of the grave. And so I took a good look around all the numerous tombs that stand beside the Agrigentine Gate. Finally I noted a little column just visible above the scrub: it was surmounted by a sphere and a cylinder. I immediately said to the Syracusans, some of whose leading citizens were with me at the time, that I believed this was the very object I had been looking for. Men were sent in with sickles to clear the site, and when the path to the monument had been opened we walked right up to it. And the verses were still visible, though approximately the second half of each line had been worn away. So one of the most famous cities in the Greek world, and in former days a great center of learning as well, would have remained in total ignorance of the tomb of the most brilliant citizen it had ever produced, had a man from Arpinum not come and pointed it out!

Cicero did not exaggerate in describing Archimedes' greatness. In fact, I have deliberately put the bar for the title of "magician" so high that progressing from the giant Archimedes, we have to leap forward no fewer than about eighteen centuries before encountering a man of similar stature. Unlike Archimedes, who said he could move the Earth, this magician insisted that the Earth was already moving.

Archimedes' Best Student

Galileo Galilei (figure 16) was born in Pisa on February 15, 1564. His father, Vincenzo, was a musician, and his mother, Giulia Ammannati, was a witty, if rather ill-disposed woman who couldn't tolerate stupidity. In 1581, Galileo followed his father's advice and enrolled in the faculty of arts of the University of Pisa to study medicine. His

Figure 16

interest in medicine, however, withered almost as soon as he got in, in favor of mathematics. Consequently, during the summer vacation of 1583, Galileo persuaded the mathematician of the Tuscan Court, Ostilio Ricci (1540–1603), to meet with his father and to convince the latter that Galileo was destined to become a mathematician. The question was indeed settled soon thereafter—the enthusiastic youth became absolutely bewitched by the works of Archimedes: "Those who read his works," he wrote, "realize only too clearly how inferior are all other minds compared with Archimedes', and what small hope is left of ever discovering things similar to the ones he discovered." At the time, little did Galileo know that he himself possessed one of those few minds that were not inferior to that of the Greek master. Inspired by the legendary story of Archimedes and the king's wreath, Galileo published in 1586 a small book entitled *The Little Balance,* about a hydrostatic balance he had invented. He later made further reference to Archimedes in a literary lecture at the Florence Academy, in which he discussed a rather unusual topic—the location and size of hell in Dante's epic poem *Inferno.*

In 1589 Galileo was appointed to the chair of mathematics at the University of Pisa, partly because of the strong recommendation of Christopher Clavius (1538–1612), a respected mathematician and astronomer from Rome, whom Galileo visited in 1587. The young mathematician's star was now definitely on the rise. Galileo spent the next three years setting forth his first thoughts on the theory of motion. These essays, which were clearly stimulated by Archimedes' work, contain a fascinating mixture of interesting ideas and false assertions. For instance, together with the pioneering realization that one can test theories about falling bodies using an inclined plane to slow down the motion, Galileo incorrectly states that when bodies are dropped from towers, " wood moves more swiftly than lead at the beginning of its motion." Galileo's inclinations and general thought process during this phase of his life have been somewhat misrepresented by his first biographer, Vincenzio Viviani (1622–1703). Viviani created the popular image of a meticulous, hard-nosed experimentalist who gained new insights exclusively from detailed observations of natural phenomena. In fact, until 1592 when he moved to Padua, Galileo's orientation and methodology were primarily mathematical. He relied mostly on thought experiments and on an Archimedean description of the world in terms of geometrical figures that obeyed mathematical laws. His chief complaint against Aristotle at that time was that the latter "was ignorant not only of the profound and more abstruse discoveries of geometry, but even of the most elementary principles of this science." Galileo also thought that Aristotle relied too heavily on sensory experiences, "because they offer at first sight some appearance of truth." Instead, Galileo proposed "to employ reasoning at all times rather than examples (for we seek the causes of effects, and these are not revealed by experience)."

Galileo's father died in 1591, prompting the young man, who had now to support his family, to take an appointment in Padua, where his salary was tripled. The next eighteen years proved to be the happiest in Galileo's life. In Padua he also began his long-term relationship with Marina Gamba, whom he never married, but who bore him three children— Virginia, Livia, and Vincenzio.

On August 4, 1597, Galileo wrote a personal letter to the great German astronomer Johannes Kepler in which he admitted that he had been a Copernican "for a long time," adding that he found in the Copernican heliocentric model a way to explain a number of natural events that could not be explained by the geocentric doctrine. He lamented the fact, however, that Copernicus "appeared to be ridiculed and hissed off the stage." This letter marked the widening of the momentous rift between Galileo and the Aristotelian cosmology. Modern astrophysics was starting to take shape.

The Celestial Messenger

On the evening of October 9, 1604, astronomers in Verona, Rome, and Padua were startled to discover a new star that rapidly became brighter than all the stars in the sky. The meteorologist Jan Brunowski, an imperial official in Prague, also saw it on October 10, and in acute agitation he immediately informed Kepler. Clouds prevented Kepler from observing the star until October 17, but once he started, he continued to record his observations for a period of about a year, and he eventually published a book about the "new star" in 1606. Today we know that the 1604 celestial spectacle did not mark the birth of a new star, but rather the explosive death of an old one. This event, now called *Kepler's supernova,* caused quite a sensation in Padua. Galileo managed to see the new star with his own eyes late in October 1604, and the following December and January he gave three public lectures on the subject to large audiences. Appealing to knowledge over superstition, Galileo showed that the absence of any apparent shift (*parallax*) in the new star's position (against the background of the fixed stars) demonstrated that the new star had to be located beyond the lunar region. The significance of this observation was enormous. In the Aristotelian world, all changes in the heavens were restricted to this side of the Moon, while the far more distant sphere of the fixed stars was assumed to be inviolable and immune to change.

The shattering of the immutable sphere had started already in 1572, when the Danish astronomer Tycho Brahe (1546–1601) observed another stellar explosion now known as *Tycho's supernova.* The 1604 event put yet another nail in the coffin of Aristotle's cosmology. But the true breakthrough in the understanding of the cosmos didn't descend from either the realm of theoretical speculation or from naked-eye observations. It was rather the outcome of simple experimentation with convex (bulging outward) and concave (curving inward) glass lenses—hold the right two of those some thirteen inches apart and distant objects suddenly appear closer. By 1608, such spyglasses started to appear all over Europe, and one Dutch and two Flemish spectacle makers even applied for patents on them. Rumors of the miraculous instrument reached the Venetian theologian Paolo Sarpi, who informed Galileo around May of 1609. Anxious to confirm the information, Sarpi also wrote to a friend in Paris, Jacques Badovere, to inquire whether the rumors were true. According to his own testimony, Galileo was "seized with a desire for the beautiful thing." He later described these events in his book *The Sidereal Messenger,* which appeared in March 1610:

> About 10 months ago a report reached my ears that a certain Fleming had constructed a spyglass by means of which visible objects, though very distant from the eye of the observer, were distinctly seen as if nearby. Of this truly remarkable effect several experiences were related, to which some persons gave credence while others denied them. A few days later the report was confirmed to me in a letter from a noble Frenchman at Paris, Jacques Badovere, which caused me to apply myself wholeheartedly to investigate means by which I might arrive at the invention of a similar instrument. This I did soon afterwards, my basis being the doctrine of refraction.

Galileo demonstrates here the same type of creatively practical thinking that characterized Archimedes—once he knew that a telescope could be built, it didn't take him long to figure out how to build one himself. Moreover, between August 1609 and March 1610, Galileo used his inventiveness to improve his telescope from a device that brought objects eight times closer to an instrument with a power of twenty. This was a considerable technical feat in itself, but Galileo's greatness was about to be revealed not in his practical know-how, but in the use to which he put his vision-enhancing tube (which he called a *perspicillum*). Instead of spying on distant ships from Venice's harbor, or examining the rooftops of Padua, Galileo pointed his telescope to the heavens. What followed was something unprecedented in scientific history. As the historian of science Noel Swerdlow puts it: "In about two months, December and January [1609 and 1610, respectively], he made more discoveries that changed the world than anyone has ever made before or since." In fact, the year 2009 has been named the International Year of Astronomy to mark the four hundredth anniversary of Galileo's first observations. What did Galileo actually do to become such a larger-than-life scientific hero? Here are only a few of his surprising achievements with the telescope.

Turning his telescope to the Moon and examining in particular the terminator—the line dividing the dark and illuminated parts—Galileo found that this celestial body had a rough surface, with mountains, craters, and vast plains. He watched how bright points of light appeared in the side veiled in darkness, and how these pinpoints widened and spread just like the light of the rising sun catching on mountaintops. He even used the geometry of the illumination to determine the height of one mountain, which turned out to be more than four miles. But this was not all. Galileo saw that the dark part of the Moon (in its crescent phase) is also faintly illuminated, and he concluded that this was due to reflected sunlight from the Earth. Just as the Earth is lit by the full Moon, Galileo asserted, the lunar surface bathes in reflected light from Earth.

While some of these discoveries were not entirely new, the strength of Galileo's evidence raised the argument to a whole new level. Until Galileo's time, there was a clear distinction between the terrestrial and the celestial, the earthly and the heavenly. The difference was not just scientific or philosophical. A rich tapestry of mythology, religion, romantic poetry, and aesthetic sensibility had been woven around the perceived dissimilarity between heaven and Earth. Now Galileo was saying something that was considered quite inconceivable. Contrary to the Aristotelian doctrine, Galileo put the Earth and a heavenly body (the Moon) on very similar footing—both had solid, rugged surfaces, and both reflected light from the Sun.

Moving yet farther from the Moon, Galileo started to observe the planets—the name coined by the Greeks for those "wanderers" in the night sky. Directing his telescope to Jupiter on January 7, 1610, he was astonished to discover three new stars in a straight line crossing the planet, two to its east and one to the west. The new stars appeared to change their positions relative to Jupiter on the following nights. On January 13, he observed a fourth such star. Within about a week from the initial discovery, Galileo reached a startling conclusion—the new stars were actually satellites orbiting Jupiter, just as the Moon was orbiting the Earth.

One of the distinguishing characteristics of the individuals who had a significant impact on the history of science was their ability to grasp immediately which discoveries were truly likely to make a difference. Another trait of many influential scientists was their skill in making the discoveries intelligible to others. Galileo was a master in both of these departments. Concerned that someone else might also discover the Jovian satellites, Galileo rushed to publish his results, and by the spring of 1610 his treatise *Sidereus Nuncius* (*The Sidereal Messenger*) appeared in Venice. Still politically astute at that point in his life, Galileo dedicated the book to the grand duke of Tuscany, Cosimo II de Medici, and he named the satellites the "Medicean Stars." Two years later, following what he referred to as an "Atlantic labor," Galileo was able to determine the orbital periods—the time it took each of the four satellites to revolve around Jupiter—to within an accuracy of a few minutes. *The Sidereal Messenger* became an instant best seller—its original five hundred copies quickly sold out—making Galileo famous across the continent.

The importance of the discovery of the Jovian satellites cannot be overemphasized. Not only were these the first bodies to be added to the solar system since the observations of the ancient Greeks, but the mere existence of these satellites removed in a single stroke one of the most serious objections to Copernicanism. The Aristotelians argued that it was impossible for the Earth to orbit the Sun, since the Earth itself had the Moon orbiting it. How could the universe have two separate centers of rotation, the Sun and the Earth? Galileo's discovery unambiguously demonstrated that a planet could have satellites orbiting it while the planet itself was following its own course around the Sun.

Another important discovery that Galileo made in 1610 was that of the phases of the planet Venus. In the geocentric doctrine, Venus was assumed to move in a small circle (an *epicycle*) superimposed on its orbit around the Earth. The center of the epicycle was supposed to always lie on the line joining the Earth and the Sun (as in figure 17a; not drawn to scale). In this case, when observed from Earth, one would expect Venus always to appear as a crescent of somewhat varying width. In the Copernican system, on the other hand, Venus's appearance should change from a small bright disk when the planet is on the other side of the Sun (as seen from Earth), to a large and almost dark disk when Venus is on the same side as Earth (figure 17b). Between those two positions Venus should pass through an entire sequence of phases similar to that of the Moon. Galileo corresponded about this important difference between the predictions of

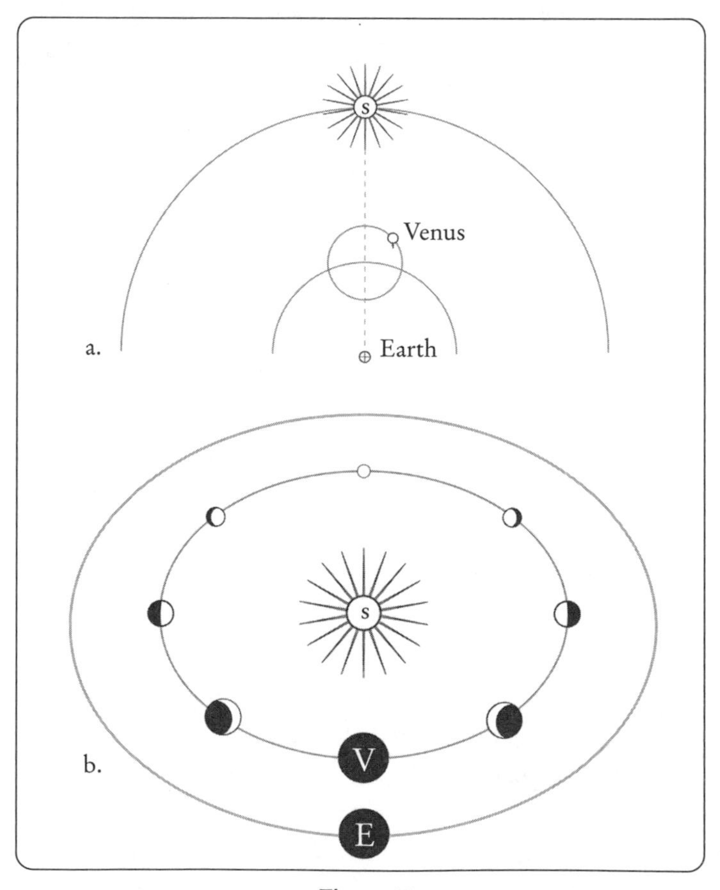

Figure 17

the two doctrines with his former student Benedetto Castelli (1578–1643), and he conducted the crucial observations between October and December of 1610. The verdict was clear. The observations confirmed conclusively the Copernican prediction, proving that Venus indeed orbits the Sun. On December 11, a playful Galileo sent Kepler the obscure anagram "*Haec immatura a me iam frustra leguntur oy*" ("This was already tried by me in vain too early"). Kepler tried unsuccessfully to decipher the hidden message and eventually gave up. In his following letter, of January 1, 1611, Galileo finally transposed the letters in the anagram to read: "*Cynthiae figuras aemulatur mater amorum*" ("The mother of love [Venus] emulates the figures of Cynthia [the Moon]").

All the findings I have described so far concerned either planets in the solar system—celestial bodies that orbit the Sun and reflect its light—or satellites revolving around these planets. Galileo also made two very significant discoveries related to stars—heavenly objects that generate their own light, such as the Sun. First, he performed observations of the Sun itself. In the Aristotelian worldview, the Sun was supposed to symbolize otherworldly perfection and immutability. Imagine the shock caused by the realization that the solar surface is far from perfect. It contains blemishes and dark spots that appear and disappear as the Sun rotates about its axis. Figure 18 shows Galileo's hand-drawn images of sunspots, about which Galileo's colleague Federico Cesi (1585–1630) wrote that they "delight both by the wonder of the spectacle and the accuracy of expression." Actually, Galileo was neither the first to see sunspots nor even the first to write about them. One pamphlet in particular, *Three Letters on Sunspots,* written by the Jesuit priest and scientist Christopher Scheiner (1573–1650) annoyed Galileo so much that he felt compelled to publish an articulate reply. Scheiner argued that it was impossible for the spots to be right on the Sun's surface. His claim was based partly on the spots being, in his opinion, too dark (he thought that they were darker than the dark parts of the Moon), and partly on the fact that they did not always appear to return to the same positions. Scheiner consequently believed that these were small planets orbiting the Sun. In his *History and Demonstrations Concerning Sunspots,* Galileo

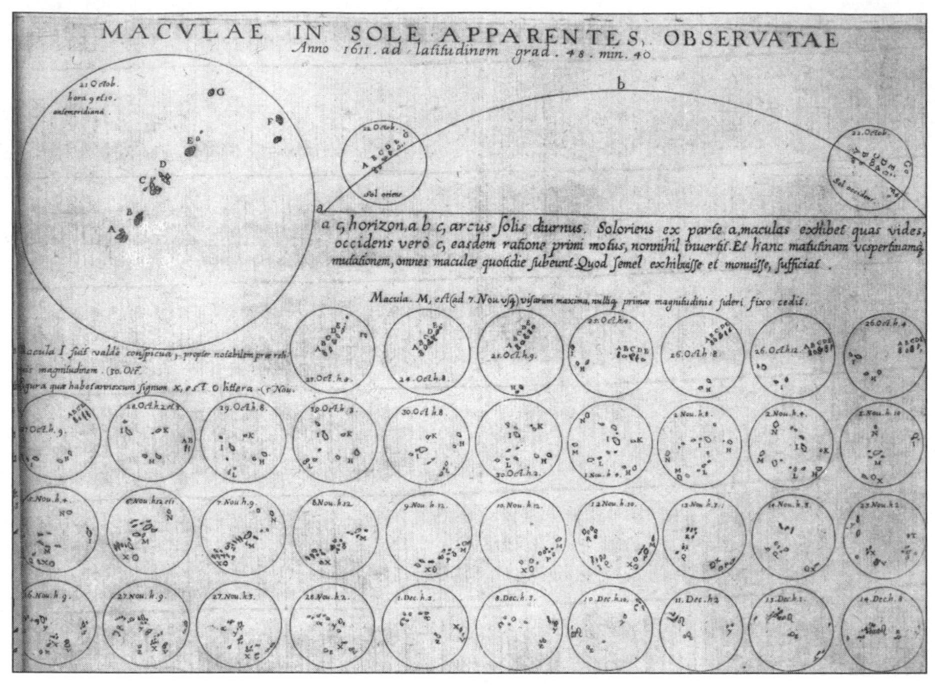

Figure 18

systematically destroyed Scheiner's arguments one by one. With a meticulousness, wit, and sarcasm that would have made Oscar Wilde jump to a standing ovation, Galileo showed that the spots were in fact not dark at all, only dark relative to the bright solar surface. In addition, Galileo's work left no doubt that the spots were right on the Sun's surface (I shall return to Galileo's demonstration of this fact later in this chapter).

Galileo's observations of other stars were truly the first human ventures into the cosmos that lies beyond our solar system. Unlike his experience with the Moon and the planets, Galileo discovered that his telescope hardly enlarged the images of stars at all. The implication was clear—stars were far more distant than planets. This was a surprise in itself, but what was truly eyepopping was the sheer *number* of new, faint stars that the telescope had revealed. In one small area around the constellation Orion alone, Galileo discovered no fewer than five hundred new stars. When Galileo turned his telescope to traverse the Milky Way—that patch of dim light that crosses the night sky—he was in for the biggest surprise yet. Even the smooth-looking bright splash broke into a countless number of stars no human had ever seen as such before. The universe suddenly got much bigger. In the somewhat dispassionate language of a scientist, Galileo reported:

> What was observed by us in the third place is the nature of matter of the Milky Way itself, which, with the aid of the spyglass, may be observed so well that all the disputes that for so many generations have vexed philosophers are destroyed by visible certainty, and we are liberated from worldly arguments. For the Galaxy is nothing else than a congeries of innumerable stars distributed in clusters. To whatever region of it you direct your spyglass, an immense number of stars immediately offer themselves to view. Of which very many appear rather large and very conspicuous but the multitude of small ones is truly unfathomable.

Some of Galileo's contemporaries reacted enthusiastically. His discoveries ignited the imagination of scientists and non-scientists alike all over Europe. The Scottish poet Thomas Seggett raved:

> *Columbus gave man lands to conquer by bloodshed,*
> *Galileo new worlds harmful to none.*
> *Which is better?*

Sir Henry Wotton, an English diplomat in Venice, managed to get hold of a copy of the *Sidereus Nuncius* the day that the book appeared. He immediately forwarded it to King James I of England, accompanied by a letter that read in part:

> I send herewith unto his Majesty the strangest piece of news (as I may justly call it) that he hath ever yet received from my part of the world; which is the annexed book (come abroad this very day) of the Mathematical Professor of Padua, who by the help of an optical instrument . . . hath discovered four new planets rolling about the sphere of Jupiter, besides many other unknown fixed stars.

Entire volumes can be written (and indeed have been written) about all of Galileo's achievements, but these lie beyond the scope of the present book. Here I only want to examine the effect that some of these astounding revelations had on Galileo's views of the universe. In particular, what relation, if any, did he perceive between mathematics and the vast, unfolding cosmos?

The Grand Book of Nature

The philosopher of science Alexandre Koyré (1892–1964) remarked once that Galileo's revolution in scientific thinking can be distilled to one essential element: the discovery that mathematics is the grammar of science. While the Aristotelians were happy with a qualitative description of nature, and even for that they appealed to Aristotle's authority, Galileo insisted that scientists should listen to nature itself, and that the keys to deciphering the universe's parlance were mathematical relations and geometrical models. The stark differences between the two approaches are exemplified by the writings of prominent members of the two camps. Here is the Aristotelian Giorgio Coresio: "Let us conclude, therefore, that he who does not want to work in darkness must consult Aristotle, the excellent interpreter of nature." To which another Aristotelian, the Pisan philosopher Vincenzo di Grazia, adds:

> Before we consider Galileo's demonstrations, it seems necessary to prove how far from the truth are those who wish to prove natural facts by means of mathematical reasoning, among whom, if I am not mistaken, is Galileo. All the sciences and all the arts have their own principles and their own causes by means of which they demonstrate the special properties of their own object. *It follows that we are not allowed to use the principles of one science to prove the properties of another* [the emphasis is mine]. Therefore, anyone who thinks he can prove natural properties with mathematical argument is simply demented, for the two sciences are very different. The natural scientist studies natural bodies that have motion as

> their natural and proper state, but the mathematician abstracts from all motion.

This idea of hermetic compartmentalization of the branches of science was precisely the type of notion that infuriated Galileo. In the draft of his treatise on hydrostatics, *Discourse on Floating Bodies,* he introduced mathematics as a powerful engine that enables humans to truly unravel nature's secrets:

> I expect a terrible rebuke from one of my adversaries, and I can almost hear him shouting in my ears that it is one thing to deal with matters physically and quite another to do so mathematically, and that geometers should stick to their fantasies, and not get involved in philosophical matters where the conclusions are different from those in mathematics. As if truth could ever be more than one; as if geometry in our day was an obstacle to the acquisition of true philosophy; as if it were impossible to be a geometer as well as a philosopher, so that we must infer as a necessary consequence that anyone who knows geometry cannot know physics, and cannot reason about and deal with physical matters physically! Consequences no less foolish than that of a certain physician who, moved by a fit of spleen, said that the great doctor Acquapendente [the Italian anatomist Hieronymus Fabricius (1537–1619) of Acquapendente], being a famous anatomist and surgeon, should content himself to remain among his scalpels and ointments without trying to effect cures by medicine, as if knowledge of surgery was opposed to medicine and destroyed it.

A simple example of how these different attitudes toward observational findings could completely alter the interpretation of natural phenomena is provided by the discovery of sunspots. As I noted earlier, the Jesuit astronomer Christopher Scheiner observed these spots competently and carefully. However, he made the mistake of allowing his Aristotelian prejudices of a perfect heaven to color his judgment. Consequently, when he discovered that the spots did not return to the same position and order, he was quick to announce that he could "free the Sun from the injury of spots." His premise of celestial immutability constrained his imagination and prevented him from considering the possibility that the spots could change, even beyond recognition. He therefore concluded that the spots *had* to be stars orbiting the Sun. Galileo's course of attack on the question of the distance of the spots from the Sun's surface was entirely different. He identified three observations that needed an explanation: First, the spots appeared to be thinner when they were near the edge of the solar disk than when they were near the disk's center. Second, the separations between the spots appeared to increase as the spots approached the center of the disk. Finally, the spots appeared to travel faster near the center than close to the edge. Galileo was able to show with a single geometrical construction that the hypothesis—that the spots were contiguous to the surface of the Sun and were carried around by it—was consistent with all the observational facts. His detailed explanation was based on the visual phenomenon of *foreshortening* on a sphere—the fact that shapes appear thinner and closer together near the edge (figure 19 demonstrates the effect for circles drawn on a spherical surface).

The importance of Galileo's demonstration for the foundations of the scientific process was tremendous. He showed that observational data become meaningful descriptions of reality only when embedded in an appropriate mathematical theory. The same observations could

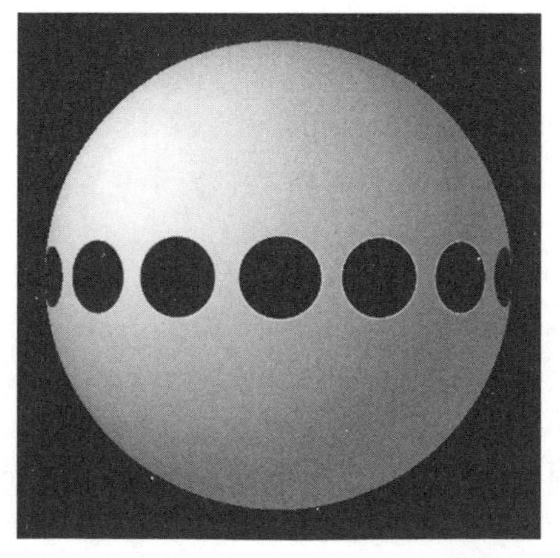

Figure 19

lead to ambiguous interpretations unless understood in a broader theoretical context.

Galileo never gave up an opportunity for a good fight. His most articulate exposition of his thoughts on the nature of mathematics and of its role in science appears in another polemic publication—*The Assayer*. This brilliant, masterfully written treatise became so popular that Pope Urban VIII had pages from it read to him during his meals. Oddly enough, Galileo's central thesis in *The Assayer* was patently wrong. He tried to argue that comets were really phenomena caused by some quirks of optical refraction on this side of the Moon.

The entire story of *The Assayer* sounds a bit as if it were taken from the libretto of an Italian opera. In the fall of 1618, three comets appeared in succession. The third one, in particular, remained visible for almost three months. In 1619, Horatio Grassi, a mathematician from the Jesuit Collegio Romano, anonymously published a pamphlet about his observations of these comets. Following in the footsteps of the great Danish astronomer Tycho Brahe, Grassi concluded that the comets were somewhere between the Moon and the Sun. The pamphlet might have gone unnoticed, but Galileo decided to respond, having been told that some Jesuits took Grassi's publication as a blow to Copernicanism. His reply was in the form of lectures (largely written by Galileo himself) that were delivered by Galileo's disciple Mario Guiducci. In the published version of these lectures, *Discourse on the Comets,* Galileo directly attacked Grassi and Tycho Brahe. This time it was Grassi's turn to take offense. Under the pseudonym of Lothario Sarsi, and posing as one of his own students, Grassi published an acrimonious reply, criticizing Galileo in no uncertain terms (the response was entitled *The Astronomical and Philosophical Balance, on which the opinions of Galileo Galilei regarding Comets are weighed, as well as those presented in the Florentine Academy by Mario Guiduccio*). In defense of his application of Tycho's methods for determining distances, Grassi (speaking as if he were his student) argued:

> Let it be granted that my master followed Tycho. Is this such a crime? Whom instead should he follow? Ptolemy [the Alexandrian originator of the heliocentric system]? Whose follow-

> ers' throats are threatened by the out-thrust sword of Mars now made closer. Copernicus? But he who is pious will rather call everyone away from him and will spurn and reject his recently condemned hypothesis. Therefore, Tycho remains as the only one whom we may approve of as our leader among the unknown courses of the stars.

This text beautifully demonstrates the fine line that Jesuit mathematicians had to walk at the beginning of the seventeenth century. On one hand, Grassi's criticism of Galileo was entirely justified and penetratingly insightful. On the other, by being forced not to commit to Copernicanism, Grassi had imposed upon himself a straitjacket that impaired his overall reasoning.

Galileo's friends were so concerned that Grassi's attack would undermine Galileo's authority that they urged the master to respond. This led to the publication of *The Assayer* in 1623 (the full title explained that in the document "are weighed with a fine and accurate balance the contents of the *Astronomical and Philosophical Weighing Scales* of Lothario Sarsi of Siguenza").

As I noted above, *The Assayer* contains Galileo's clearest and most powerful statement concerning the relation between mathematics and the cosmos. Here is that remarkable text:

> I believe Sarsi is firmly convinced that it is essential in philosophy to support oneself by the opinion of some famous author, as if when our minds are not wedded to the reasoning of someone else they ought to remain completely barren and sterile. Perhaps he thinks that philosophy is a book of fiction created by some man, like the *Iliad* or *Orlando Furioso* [an epic sixteenth century poem by Ludovico Ariosto]—books in which the least important thing is whether what is written in them is true. Sig. Sarsi, this is not how matters stand. *Philosophy is written in that great book which ever lies before our eyes (I mean the universe) but we cannot understand it if we do not first learn the language and grasp the characters in which it is written. It is written in the language of mathematics, and the*

> *characters are triangles, circles and other geometrical figures, without which it is humanly impossible to comprehend a single word of it, and without which one wanders in vain through a dark labyrinth.* [emphasis added]

Mind-boggling, isn't it? Centuries before the question of why mathematics was so effective in explaining nature was even asked, Galileo thought he already knew the answer! To him, mathematics was simply the language of the universe. To understand the universe, he argued, one must speak this language. God is indeed a mathematician.

The full range of ideas in Galileo's writings paints an even more detailed picture of his views on mathematics. First, we must realize that to Galileo, mathematics ultimately meant geometry. Rarely was he interested in measuring values in absolute numbers. He described phenomena mainly with proportions among quantities and in relative terms. In this again, Galileo was a true disciple of Archimedes, whose principle of the lever and method of comparative geometry he used effectively and extensively. A second interesting point, which is revealed especially in Galileo's last book, is the distinction he makes between the roles of geometry and logic. The book itself, *Discourses and Mathematical Demonstrations Concerning Two New Sciences,* is written in the form of lively discussions among three interlocutors, Salviati, Sagredo, and Simplicio, whose roles are quite clearly demarcated. Salviati is effectively Galileo's spokesman. Sagredo, the aristocratic philosophy lover, is a man whose mind has already escaped from the illusions of Aristotelian common sense and who can therefore be persuaded by the strength of the new mathematical science. Simplicio, who in Galileo's previous work was portrayed as being under the spell of Aristotelian authority, appears here as an open-minded scholar. On the second day of the argument, Sagredo has an interesting exchange with Simplicio:

> Sagredo: What shall we say, Simplicio? Must we not confess that the power of geometry is the most potent instrument of all to sharpen the mind and dispose it to reason perfectly, and to speculate? Didn't Plato have good reason to want his pupils to be first grounded in mathematics?

Simplicio appears to agree and he introduces the comparison with logic:

> Simplicio: Truly I begin to understand that although logic is a very excellent instrument to govern our reasoning, it does not compare with the sharpness of geometry in awakening the mind to discovery.

Sagredo then sharpens the distinction:

> Sagredo: It seems to me that logic teaches how to know whether or not reasoning and demonstrations already discovered are conclusive, but I do not believe that it teaches how to find conclusive reasoning and demonstrations.

Galileo's message here is simple—he believed that geometry was the tool by which new truths are *discovered*. Logic, on the other hand, was to him the means by which discoveries already made are *evaluated and critiqued.* In chapter 7 we shall examine a different perspective, according to which the whole of mathematics stems from logic.

How did Galileo arrive at the idea that mathematics was nature's language? After all, a philosophical conclusion of this magnitude could not have suddenly materialized out of thin air. Indeed, the roots of this conception can be traced all the way back to the writings of Archimedes. The Greek master was the first to use mathematics to explain natural phenomena. Then, via a tortuous path passing through some medieval calculators and Italian court mathematicians, the nature of mathematics gained the status of a subject worthy of discussion. Eventually, some of the Jesuit mathematicians of Galileo's time, Christopher Clavius in particular, also acknowledged the fact that mathematics might occupy some middle ground between metaphysics—the philosophical principles of the nature of being—and physical reality. In the preface ("Prolegomena") to his *Comments on Euclid's "Elements,"* Clavius wrote:

> Since the mathematical disciplines deal with things which are considered apart from any sensible matter, although they are

> immersed in material things, it is clear that they hold a place intermediate between metaphysics and natural science, if we consider their subject matter.

Galileo was not satisfied with mathematics as the mere go-between or conduit. He took the extra bold step of equating mathematics with God's native tongue. This identification, however, raised another serious problem—one that was about to have a dramatic impact on Galileo's life.

Science and Theology

According to Galileo, God spoke in the language of mathematics in designing nature. According to the Catholic Church, God was the "author" of the Bible. What was one to make then of those cases where the mathematically based scientific explanations seemed to contradict the scriptures? The theologians of the 1546 Council of Trent answered in no uncertain terms: "No one relying on his own judgment and distorting the Sacred Scriptures according to his own conception shall dare to interpret them contrary to that sense which Holy Mother Church, to whom it belongs to judge of their true sense and meaning, has held or does hold." Accordingly, when in 1616 theologians were asked to give their opinion on Copernicus's heliocentric cosmology, they concluded that it was "formally heretical, since it explicitly contradicts in many places the sense of the Holy Scripture." In other words, what was truly at the heart of the Church's objection to Galileo's Copernicanism was not so much the removal of the Earth from its central position in the cosmos, but rather the challenge to the church's authority in interpreting the scriptures. In a climate in which the Roman Catholic Church was already feeling embattled by controversies with Reformation theologians, Galileo and the Church were on a clear collision course.

Events started to unfold rapidly toward the end of 1613. Galileo's former student, Benedetto Castelli, made a presentation of the new astronomical discoveries to the grand duke of Tuscany and his entourage. Predictably, he was pressured to explain the apparent discrepancies between the Copernican cosmology and some biblical accounts, such as the one in which God stopped the Sun and the Moon in their courses to allow Joshua and the Israelites to complete their victory over the Emorites in the Ayalon Valley. Even though Castelli reported that he "behaved like a champion" in defending Copernicanism, Galileo was somewhat disturbed by the news of this confrontation, and he felt compelled to express his own views about contradictions between science and the Holy Scriptures. In a long letter to Castelli dated December 21, 1613, Galileo writes:

> It was necessary, however in the Holy Scripture, in order to accommodate itself to the understanding of the majority, to say many things which apparently differ from the precise meaning. Nature, on the contrary, is inexorable and unchangeable, and cares not whether her hidden causes and modes of working are intelligible to the human understanding or not, and never deviates on that account from the prescribed laws. It appears to me therefore that no effect of nature, which experience places before our eyes, or is the necessary conclusion derived from evidence, should be rendered doubtful by passages of Scripture which contain thousands of words admitting of various interpretations, for every sentence of Scripture is not bound by such rigid laws as is every effect of nature.

This interpretation of the biblical meaning was clearly at odds with that of some of the more stringent theologians. For instance, the Dominican Domingo Bañez wrote in 1584: "The Holy Spirit not only inspired all that is contained in the Scripture, he also dictated and suggested every word with which it was written." Galileo was obviously not convinced. In his *Letter to Castelli* he added:

> I am inclined to think that the authority of Holy Scripture is intended to convince men of those truths which are necessary for their salvation, and which being far above man's understanding cannot be made credible by any learning, or any other means than revelation by the Holy Spirit. But that the same

> God that has endowed us with senses, reason, and understanding, does not permit us to use them, and desires to acquaint us in any other way with such knowledge as we are in a position to acquire for ourselves by means of those faculties, *that* it seems to me I am not bound to believe, especially concerning those sciences about which the Holy Scripture contains only small fragments and varying conclusions; and this is precisely the case with astronomy, of which there is so little that the planets are not even all enumerated.

A copy of Galileo's letter made it to the Congregation of the Holy Office in Rome, where affairs concerning faith were commonly evaluated, and especially to the influential Cardinal Robert Bellarmine (1542–1621). Bellarmine's original reaction to Copernicanism was rather moderate, since he regarded the entire heliocentric model as "a way to save the appearances, in the manner of those who have proposed epicycles but do not really believe in their existence." Like others before him, Bellarmine too treated the mathematical models put forth by astronomers as merely convenient gimmicks, designed to describe what humans observed, without being anchored in any physical reality. Such "saving the appearances" devices, he argued, do not demonstrate that the Earth is really moving. Consequently, Bellarmine saw no immediate threat from Copernicus's book (*De Revolutionibus*), even though he was quick to add that to claim that the Earth was moving would not only "irritate all scholastic philosophers and theologians" but would also "harm the Holy Faith by rendering Holy Scripture false."

The full details of the rest of this tragic story are beyond the scope and main focus of the present book, so I'll describe them only briefly here. The Congregation of the Index banned Copernicus's book in 1616. Galileo's further attempts to rely on numerous passages from the most revered of the early theologians—St. Augustine—to support his interpretation of the relation between the natural sciences and Scripture did not gain him much sympathy. In spite of articulate letters in which his main thesis was that there is no disagreement (other than superficial) between the Copernican theory and the biblical texts, the theologians of his day regarded Galileo's arguments as an uninvited foray into their domain. Cynically, these same theologians did not hesitate to express opinions on scientific matters.

As the dark clouds were gathering on the horizon, Galileo continued to believe that reason would prevail—a huge mistake when it comes to challenging faith. Galileo published his *Dialogue Concerning the Two Chief World Systems* in February of 1632 (figure 20 shows the frontispiece of the first edition). This polemical text was Galileo's most detailed exposition of his Copernican ideas. Moreover, Galileo argued that by pursuing science using the language of mechanical equilibrium and mathematics, humans could understand the divine mind. Put differently, when a person finds a solution to a problem using proportional geometry, the insights and understanding gained are godlike. The church's reaction was swift and decisive. The circulation of the *Dialogue* was forbidden as early as August of the year of its publication. In the following month, Galileo was summoned to Rome to defend himself against the charges of heresy. Galileo was brought to trial on April 12, 1633, and he was found "vehemently suspect of heresy" on June 22, 1633. The judges accused Galileo "of having believed and held the doctrine—which is false and contrary to the sacred and divine Scriptures—that the Sun is the center of the world and does not move from east to west and that the Earth moves and is not the center of the world." The sentence was harsh:

> We condemn you to the formal prison of this Holy Office during our pleasure, and by way of salutary penance we enjoin that for three years to come you repeat once a week the seven penitential Psalms. Reserving to ourselves liberty to moderate, commute, or take off, in whole or in part, the aforementioned penalties and penance.

The devastated seventy-year-old Galileo could not withstand the pressure. His spirit broken, Galileo submitted his letter of abjuration, in which he committed to "abandon completely the false opinion that the Sun is at the center of the world and does not move and that the Earth is not the center of the world and moves." He concluded:

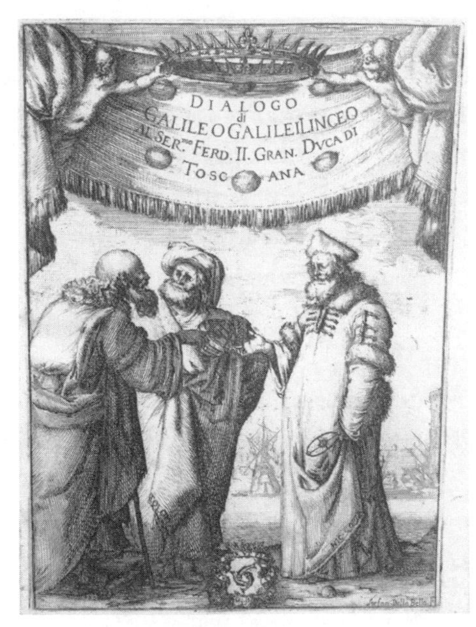

Figure 20

> Therefore, desiring to remove from the minds of your Eminences, and of all faithful Christians, this vehement suspicion justly conceived against me, with sincere heart and unfeigned faith I abjure, curse, and detest the aforesaid errors and heresies and generally every other error, heresy, and sect whatsoever contrary to the Holy Church, and I swear that in future I will never again say or assert, verbally or in writing, anything that might furnish occasion for a similar suspicion regarding me.

Galileo's last book, *Discourses and Mathematical Demonstrations Concerning Two New Sciences,* was published in July 1638. The manuscript was smuggled out of Italy and was published in Leiden in Holland. The content of this book truly and powerfully expressed the sentiment embodied in the legendary words "*Eppur si muove*" ("And yet it moves"). That defiant phrase, commonly put in Galileo's mouth at the end of his trial, was probably never uttered.

On October 31, 1992, the Catholic Church finally decided to "rehabilitate" Galileo. Recognizing that Galileo was right all along, but still avoiding direct criticism of the Inquisition, Pope John Paul II said:

> Paradoxically, Galileo, a sincere believer, proved himself more perspicacious on this issue [apparent discrepancies between science and the scriptures] than his theologian adversaries. The majority of theologians did not perceive the formal distinction that exists between the Holy Scripture in itself and its interpretation, and this led them unduly transferring to the field of religious doctrine an issue which actually belongs to scientific research.

Newspapers around the world had a feast. The *Los Angeles Times* declared: "It's Official: The Earth Revolves Around the Sun, Even for the Vatican."

Many were not amused. Some saw this *mea culpa* by the church as far too little, far too late. The Spanish Galileo scholar Antonio Beltrán Marí noted:

> The fact that the Pope continues to consider himself an authority capable of saying something relevant about Galileo and his science shows that, on the Pope's side, nothing has changed. He is behaving in exactly the same manner as Galileo's judges, whose mistake he now recognizes.

To be fair, the Pope found himself in a no-win situation. Any decision on his part, whether to ignore the issue and keep Galileo's condemnation on the books, or to finally acknowledge the church's error, was likely to be criticized. Still, at a time when there are attempts to introduce biblical creationism as an alternative "scientific" theory (under the thinly veiled title of "intelligent design"), it is good to remember that Galileo already fought this battle almost four hundred years ago—and won!

CHAPTER 4

MAGICIANS: THE SKEPTIC AND THE GIANT

In one of the seven skits in the movie *Everything You Always Wanted to Know About Sex* (*But Were Afraid to Ask)*, Woody Allen plays a court jester who does comic routines for a medieval king and his entourage. The jester has the hots for the queen, so he gives her an aphrodisiac, hoping to seduce her. The queen does become attracted to the jester, but alas, she has a huge padlock on her chastity belt. Faced with this frustrating situation in the queen's bedroom, the jester utters nervously: "I must think of something quickly, before the Renaissance will be here and we will *all* be painting."

Jokes aside, this exaggeration is an understandable description of the events in Europe during the fifteenth and sixteenth centuries. The Renaissance has indeed produced such a wealth of masterpieces in painting, sculpture, and architecture that to this very day, these astonishing works of art form a major part of our culture. In science, the Renaissance witnessed the heliocentric revolution in astronomy, led by Copernicus, Kepler, and especially Galileo. The new view of the universe afforded by Galileo's observations with the telescope, and the insights gained from his experiments in mechanics, perhaps more than anything else motivated the mathematical developments of the following century. Amidst the first signs of crumbling of the Aristotelian philosophy and the challenges to the Church's theological ideology, philosophers started to search for a new foundation on which to erect human knowledge. Mathematics, with its seemingly certain body of truths, provided what appeared to be the soundest base for a new start.

The man who embarked on the rather ambitious task of discovering a formula that would somehow discipline all rational thought and unify all knowledge, science, and ethics was a young French officer and gentleman named René Descartes.

A Dreamer

Many regard Descartes (figure 21) as both the first great modern philosopher and the first modern biologist. When you add to these impressive credentials the fact that the English empiricist philosopher John Stuart Mill (1806–73) characterized one of Descartes' achievements in mathematics as "the greatest single step ever made in the progress of the exact sciences," you begin to realize the immensity of Descartes' power of intellect.

René Descartes was born on March 31, 1596, at La Haye, France. In honor of its most celebrated resident, the town was renamed La Haye–Descartes in 1801, and since 1967 it is known simply as Descartes. At the age of eight, Descartes entered the Jesuit College of La Flèche, where he studied Latin, mathematics, classics, science, and scholastic philosophy until 1612. Because of his relatively fragile health, Descartes was excused from having to get up at the brutal hour of five a.m., and he was allowed to spend the morning hours in bed. Later in life, he continued to use the early part of the day for contemplation, and he once told the French mathematician Blaise Pascal that the only way for him to stay healthy and be productive was to never get up before he felt comfortable doing so. As we shall soon see, this statement turned out to be tragically prophetic.

After La Flèche, Descartes graduated from the University of Poitiers as a lawyer, but he never actually practiced law. Restless and eager to see the world, Descartes decided to join the army of Prince Maurice of Orange, which was then stationed at Breda in the United Provinces (The Netherlands). An accidental encounter in Breda was to become very significant in Descartes' intellectual development.

Figure 21

According to the traditional story, while wandering in the streets, he suddenly saw a billboard that appeared to present a challenging problem in mathematics. Descartes asked the first passer-by to translate the text for him from Dutch into either Latin or French. A few hours later, Descartes succeeded in solving the problem, thus convincing himself that he really had an aptitude for mathematics. The translator turned out to be none other than the Dutch mathematician and scientist Isaac Beeckman (1588–1637), whose influence on Descartes' physico-mathematical investigations continued for years. The next nine years saw Descartes alternating between the hurly-burly of Paris and service in the military corps of several armies. In a Europe in the throes of religious and political struggle and the commencement of the Thirty Years' War, it was relatively easy for Descartes to find battles or marching battalions to join, be it in Prague, Germany, or Transylvania. Nevertheless, throughout this period he continued, as he put it, "over head and ears in the study of mathematics."

On November 10, 1619, Descartes experienced three dreams that not only had a dramatic effect on the rest of *his* life, but which also marked perhaps the beginning of the modern world. When later describing the event, Descartes says in one of his notes: "I was filled with enthusiasm and discovered the foundations of a wonderful science." What were these influential dreams about?

Actually, two were nightmares. In the first dream, Descartes found himself caught in a turbulent whirlwind that revolved him violently on his left heel. He was also terrified by an endless sensation of falling down at each step. An old man appeared and attempted to present him with a melon from a foreign land. The second dream was yet another vision of horror. He was trapped in a room with ominous thunderclaps and sparks flying all around. In sharp contrast to the first two, the third dream was a picture of calm and meditation. As his eyes scanned the room, Descartes saw books appearing and disappearing on a table. They included an anthology of poems entitled *Corpus Poetarum* and an encyclopedia. He opened the anthology at random and caught a glimpse of the opening line of a poem by the fourth century Roman poet Ausonius. It read: "*Quod vitae sectabor iter?*" ("What road shall I pursue in life?"). A man miraculously appeared out of thin air and cited another verse: "*Est et non*" ("Yes and no" or "It is and it is not"). Descartes wanted to show him the Ausonius verse, but the entire vision disappeared into nothingness.

As is usually the case with dreams, their significance lies not so much in their actual content, which is often perplexing and bizarre, but in the interpretation the dreamer chooses to give them. In Descartes' case, the effect of these three enigmatic dreams was astounding. He took the encyclopedia to signify the collective scientific knowledge and the anthology of poetry to portray philosophy, revelation, and enthusiasm. The "Yes and no"—the famous opposites of Pythagoras—he understood as representing truth and falsehood. (Not surprisingly, some psychoanalytical interpretations suggested sexual connotations in relation to the melon.) Descartes was absolutely convinced that the dreams pointed him in the direction of the unification of the whole of human knowledge by the means of reason. He resigned from the army in 1621 but continued to travel and study mathematics for the next five years. All of those who met Descartes during that time, including the influential spiritual leader Cardinal Pierre de Bérulle (1575–1629), were deeply impressed with his sharpness and clarity of thought. Many encouraged him to publish his ideas. With any other young man, such fatherly words of wisdom might have had the same counterproductive effect that the one-word career advice "Plastics!" had on Dustin Hoffman's character in the movie *The Graduate*, but Descartes was different. Since he had already committed to the goal of searching for the truth, he was easily persuaded. He moved to Holland, which at the time seemed to offer a more tranquil intellectual milieu, and for the next twenty years produced one tour de force after another.

Descartes published his first masterpiece on the foundations of science, *Discourse on the Method of Properly Guiding the Reason and Seeking for Truth in the Sciences*, in 1637 (figure 22 shows the frontispiece of the first edition). Three outstanding appendices—on optics,

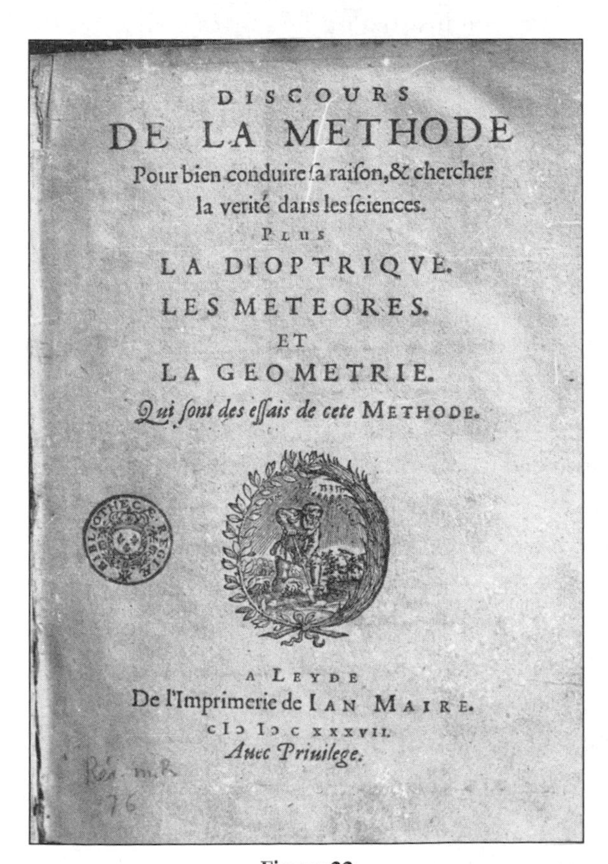

DISCOURS
DE LA METHODE
Pour bien conduire ſa raiſon, & chercher
la verité dans les ſciences.
PLUS
LA DIOPTRIQVE.
LES METEORES.
ET
LA GEOMETRIE.
Qui ſont des eſſais de cete METHODE.

A LEYDE
De l'Imprimerie de IAN MAIRE.
CIƆ IƆ C XXXVII.
Avec Privilege.

Figure 22

meteorology, and geometry—accompanied this treatise. Next came his philosophical work, *Meditations on First Philosophy,* in 1641, and his work on physics, *Principles of Philosophy,* in 1644. Descartes was by then famous all over Europe, counting among his admirers and correspondents the exiled Princess Elisabeth of Bohemia (1618–80). In 1649 Descartes was invited to instruct the colorful Queen Christina of Sweden (1626–89) in philosophy. Having always had a soft spot for royalty, Descartes agreed. In fact, his letter to the queen was so full of expressions of courtly seventeenth century awe, that today it looks utterly ridiculous: "I dare to protest here to Your Majesty that she could command nothing to me so difficult that I would not always be ready to do everything possible to execute it, and that even if I had been born a Swede or a Finn, I could not be more zealous nor more perfectly [for you] than I am." The iron-willed twenty-three-year-old queen insisted on Descartes giving her the lessons at the ungodly hour of five o'clock in the morning. In a land that was so cold that, as Descartes wrote to his friend, even thoughts froze there, this proved to be deadly. "I am out of my element here," Descartes wrote, "and I desire only tranquility and repose, which are goods the most powerful kings on earth cannot give to those who cannot obtain them for themselves." After only a few months of braving the brutal Swedish winter in those dark morning hours that he had managed to avoid throughout his entire life, Descartes contracted pneumonia. He died at age fifty-three on February 11, 1650, at four o'clock in the morning, as if trying to avoid another wake-up call. The man whose works announced the modern era fell victim to his own snobbish tendencies and the caprices of a young queen.

Descartes was buried in Sweden, but his remains, or at least part of them, were transferred to France in 1667. There, the remains were displaced multiple times, until they were eventually buried on February 26, 1819, in one of the chapels of the Saint-Germain-des-Prés cathedral. Figure 23 shows me next to the simple black plaque celebrating Descartes. A skull claimed to be that of Descartes was passed from hand to hand in Sweden until it was bought by a chemist named Berzelius, who transported it to France. That skull is currently at the Natural Science Museum, which is part of the Musée de l'Homme

Figure 23

(the Museum of Man) in Paris. The skull is often on display opposite the skull of a Neanderthal man.

A Modern

The label "modern," when attached to a person, usually refers to those individuals who can converse comfortably with their twentieth (or by now, twenty-first) century professional peers. What makes Descartes a true modern is the fact that he dared *to question* all the philosophical and scientific assertions that were made before his time. He once noted that his education served only to advance his perplexity and to make him aware of his own ignorance. In his celebrated *Discourse* he wrote: "I observed with regard to philosophy, that despite being cultivated for many centuries by the best minds, it contained no point which was not disputed and hence doubtful." While the fate of many of Descartes' own philosophical ideas was not going to be much different in that significant shortcomings in his propositions have been pointed out by later philosophers, his fresh skepticism of even the most basic concepts certainly makes him modern to the core. More important from the perspective of the present book, Descartes recognized that the methods and reasoning process of mathematics produced precisely the kind of *certainty* that the scholastic philosophy before his time lacked. He pronounced clearly:

> Those long chains, composed of very simple and easy reasonings, which geometers customarily use to arrive at their most difficult demonstrations, gave me occasion to suppose that *all the things which fall within the scope of human knowledge are interconnected in the same way* [the emphasis is mine]. And I thought that, provided we refrain from accepting anything as true which is not, and always keep to the order required for deducing one thing from another, there can be nothing too remote to be reached in the end or too well hidden to be discovered.

This bold statement goes, in some sense, even beyond Galileo's views. It is not only the physical universe that is written in the language of mathematics; all of human knowledge follows the logic of mathematics. In Descartes' words: "It [the method of mathematics] is a more powerful instrument of knowledge than any other that has been bequeathed to us by human agency, as being the source of all others." One of Descartes' goals became, therefore, to demonstrate that the world of physics, which to him was a mathematically describable reality, could be depicted without having to rely on any of our often-misleading sensory perceptions. He advocated that the mind should filter what the eye sees and turn the perceptions into ideas. After all, Descartes argued, "there are no certain signs to distinguish between being awake and being asleep." But, Descartes wondered, if everything we perceive as reality could in fact be only a dream, how are we to know that even the Earth and the sky are not some "delusions of dreams" installed in our senses by some "malicious demon of infinite power"? Or, as Woody Allen once put it: "What if everything is an illusion and nothing exists? In that case, I definitely overpaid for my carpet."

For Descartes, this deluge of troubling doubts eventually produced what has become his most memorable argument: *Cogito ergo sum* (I am thinking, therefore I exist). In other words, behind the thoughts there must be a conscious mind. Paradoxically perhaps, the act of doubting cannot itself be doubted! Descartes attempted to use this seemingly slight beginning to construct a complete enterprise of reliable knowledge. Whether it was in philosophy, optics, mechanics, medicine, embryology, or meteorology, Descartes tried his hand at it all and achieved accomplishments of some significance in every one of these disciplines. Still, in spite of his insistence on the human capacity to reason, Descartes did not believe that logic alone could uncover fundamental truths. Reaching essentially the same conclusion as Galileo, he noted: "As for logic, its syllogisms and the majority of its other percepts are of avail rather in the communication of what we already know . . . than in the investigation of the unknown." Instead, throughout his heroic endeavor to reinvent, or establish, the foundations of entire disciplines, Descartes attempted to use the principles that he had distilled from the mathematical method to ensure that he was proceeding on solid ground. He described these rigorous guidelines in his *Rules for the Direction of the Mind*. He would start with truths about which he had no doubt (similar to the axioms in Euclid's geometry); he would attempt to break up difficult problems into more manageable ones; he would proceed from the rudimentary to the intricate; and he would double-check his entire procedure to satisfy himself that no potential solution has been ignored. Needless to say, even this carefully constructed, arduous process could not make Descartes' conclusions immune to error. In fact, even though Descartes is best known for his monumental breakthroughs in philosophy, his most enduring contributions have been in mathematics. I shall now concentrate in particular on that one brilliantly simple idea that John Stuart Mill referred to as the "greatest single step ever made in the progress of the exact sciences."

The Mathematics of a New York City Map

Take a look at the partial map of Manhattan in figure 24. If you are standing at the corner of Thirty-fourth Street and Eighth Avenue and you have to meet someone at the corner of Fifty-ninth Street and Fifth

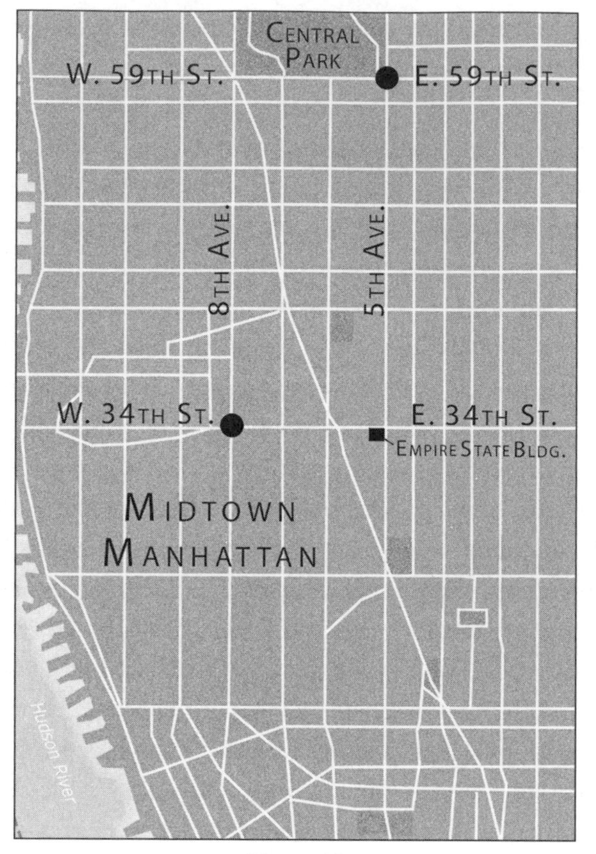

Figure 24

Avenue, you will have no trouble finding your way, right? This was the essence of Descartes' idea for a new geometry. He outlined it in a 106-page appendix entitled *La Géométrie* (*Geometry*) to his *Discourse on the Method*. Hard to believe, but this remarkably simple concept revolutionized mathematics. Descartes started with the almost trivial fact that, just as the map of Manhattan shows, a pair of numbers on the plane can determine the position of a point unambiguously (e.g., point A in figure 25a). He then used this fact to develop a powerful theory of curves—*analytical geometry.* In Descartes' honor, the pair of intersecting straight lines that give us the reference system is known as a *Cartesian coordinate system*. Traditionally, the horizontal line is labeled the "*x* axis," the vertical line the "*y* axis," and the point of intersection is known as the "origin." The point marked "A" in figure 25a, for instance, has an x coordinate of 3 and a y coordinate of 5, which is symbolically denoted by the ordered pair of numbers (3,5). (Note that the origin is designated (0,0).) Suppose now that we want to somehow characterize all the points in the plane that are at a distance of precisely 5 units from the origin. This is, of course, precisely the geometrical definition of a circle around the origin, with a radius of five units (figure 25b). If you take the point (3,4) on this circle, you find that its coordinates satisfy $3^2 + 4^2 = 5^2$. In fact, it is easy to show (using the Pythagorean theorem) that the coordinates (x, y) of any point on this circle satisfy $x^2 + y^2 = 5^2$. Furthermore, the points on the circle are the only points in the plane for whose coordinates this equation ($x^2 + y^2 = 5^2$) holds true. This means that the algebraic equation $x^2 + y^2 = 5^2$ precisely and uniquely characterizes this circle. In other words, Descartes discovered a way to represent a geometrical curve by an algebraic equation or numerically and vice versa. This may not sound exciting for a simple circle, but every graph you have ever seen, be it of the weekly ups and downs of the stock market, the temperature at the North Pole over the past century, or the rate of expansion of the universe, is based on this ingenious idea of Descartes'. Suddenly, geometry and algebra were no longer two separate branches

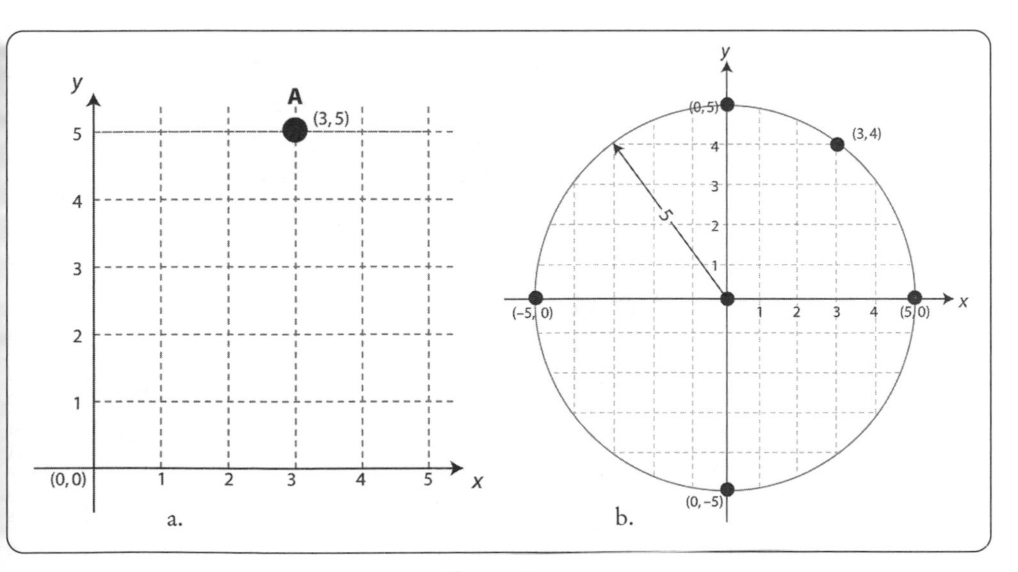

Figure 25

of mathematics, but rather two representations of the same truths. The equation describing a curve contains implicitly every imaginable property of the curve, including, for instance, all the theorems of Euclidean geometry. And this was not all. Descartes pointed out that different curves could be drawn on the same coordinate system, and that their points of intersection could be found simply by finding the solutions that are common to their respective algebraic equations. In this way, Descartes managed to exploit the strengths of algebra to correct for what he regarded as the disturbing shortcomings of classical geometry. For instance, Euclid defined a point as an entity that has no parts and no magnitude. This rather obscure definition became forever obsolete once Descartes defined a point in the plane simply as the ordered pair of numbers (x,y). But even these new insights were just the tip of the iceberg. If two quantities x and y can be related in such a way that for every value of x there corresponds a unique value of y, then they constitute what is known as a *function,* and functions are truly ubiquitous. Whether you are monitoring your daily weight while on a diet, the height of your child on consecutive birthdays, or the dependence of your car's gas mileage on the speed at which you drive, the data can all be represented by functions.

Functions are truly the bread and butter of modern scientists, statisticians, and economists. Once many repeated scientific experiments or observations produce the same functional interrelationships, those may acquire the elevated status of *laws of nature*—mathematical descriptions of a behavior all natural phenomena are found to obey. For instance, Newton's law of gravitation, to which we shall return later in this chapter, states that when the distance between two point masses is doubled, the gravitational attraction between them always decreases by a factor of four. Descartes' ideas therefore opened the door for a systematic mathematization of nearly everything—the very essence of the notion that God is a mathematician. On the purely mathematical side, by establishing the equivalence of two perspectives of mathematics (algebraic and geometric) previously considered disjoint, Descartes expanded the horizons of mathematics and paved the way to the modern arena of *analysis,* which allows mathematicians to comfortably cross from one mathematical subdiscipline into another. Consequently, not only did a variety of phenomena become describable by mathematics, but mathematics itself became broader, richer, and more unified. As the great mathematician Joseph-Louis Lagrange (1736–1813) put it: "As long as algebra and geometry traveled separate paths their advance was slow and their applications limited. But when these two sciences joined company, they drew from each other fresh vitality and thenceforward marched on at a rapid pace towards perfection."

As important as Descartes' achievements in mathematics were, he himself did not limit his scientific interests to mathematics. Science, he said, was like a tree, with metaphysics being the roots, physics the trunk, and the three main branches representing mechanics, medicine, and morals. The choice of the branches may appear somewhat surprising at first, but in fact the branches symbolized beautifully the three major areas to which Descartes wanted to apply his new ideas: the universe, the human body, and the conduct of life. Descartes spent the first four years of his stay in Holland—1629 to 1633— writing his treatise on cosmology and physics, *Le Monde* (*The World*). Just as the book was ready to go to press, however, Descartes was shocked by some troubling news. In a letter to his friend and critic, the natural philosopher Marin Mersenne (1588–1648), he lamented:

> I had intended to send you my *World* as a New Year gift, and only two weeks ago I was quite determined to send you at least a part of it, if the whole work could not be copied in time. But I have to say that in the meantime I took the trouble to inquire in Leiden and Amsterdam whether Galileo's *World System* was available, for I thought I had heard that it was published in Italy last year. I was told that it had indeed been published, but that all the copies had immediately been burnt at Rome, and that Galileo had been convicted and fined. I was so astonished at this that I almost decided to burn all my papers, or at least to let no one see them. For I could not imagine that he—an Italian, and as I understand, in the good graces of the Pope—could have been made a criminal for any other reason than that he tried, as he no doubt did, to establish that the Earth moves. I know that some Cardinals had already censured this view, but I thought I had

> heard it said that all the same it was taught publicly in Rome. *I must admit that if the view is false, so too are the entire foundations of my philosophy* [my emphasis], for it can be demonstrated from them quite clearly. And it is so closely interwoven in every part of my treatise that I could not remove it without rendering the whole work defective. But for all the world I did not want to publish a discourse in which a single word could be found that the Church would have disapproved of; so I preferred to suppress it rather than to publish it in a mutilated form.

Descartes had indeed abandoned *The World* (the incomplete manuscript was eventually published in 1664), but he did incorporate most of the results in his *Principles of Philosophy*, which appeared in 1644. In this systematic discourse, Descartes presented his laws of nature and his theory of vortices. Two of his laws closely resemble Newton's famous first and second laws of motion, but the others were in fact incorrect. The theory of vortices assumed that the Sun was at the center of a whirlpool created in the continuous cosmic matter. The planets were supposed to be swept around by this vortex like leaves in an eddy formed in the flow of a river. In turn, the planets were assumed to form their own secondary vortices that carried the satellites around. While Descartes' theory of vortices was spectacularly wrong (as Newton ruthlessly pointed out later), it was still interesting, being the first serious attempt to formulate a theory of the universe as a whole that was based on the same laws that apply on the Earth's surface. In other words, to Descartes there was no difference between terrestrial and celestial phenomena—the Earth was part of a universe that obeyed uniform physical laws. Unfortunately, Descartes ignored his own principles in constructing a detailed theory that was based neither on self-consistent mathematics nor on observations. Nevertheless, Descartes' scenario, in which the Sun and the planets somehow disturb the smooth universal matter around them, contained some elements that much later became the cornerstone of Einstein's theory of gravity. In Einstein's theory of general relativity, gravity is not some mysterious force that acts across the vast distances of space. Rather, massive bodies such as the Sun warp the space in their vicinity, just as a bowling ball would cause a trampoline to sag. The planets then simply follow the shortest possible paths in this warped space.

I have deliberately left out of this extremely brief description of Descartes' ideas almost all of his seminal work in philosophy, because this would have taken us too far afield from the focus on the nature of mathematics (I shall return to some of his thoughts about God later in the chapter). I cannot refrain, however, from including the following amusing commentary that was written by the British mathematician Walter William Rouse Ball (1850–1925) in 1908:

> As to his [Descartes'] philosophical theories, it will be sufficient to say that he discussed the same problems which have been debated for the last two thousand years, and probably will be debated with equal zeal two thousand years hence. It is hardly necessary to say that the problems themselves are of importance and interest, but from the nature of the case no solution ever offered is capable either of rigid proof or disproof; all that can be effected is to make one explanation more probable than another, and whenever a philosopher like Descartes believes that he has at last finally settled a question it has been possible for his successors to point out the fallacy in his assumptions. I have read somewhere that philosophy has always been chiefly engaged with the inter-relations of God, Nature, and Man. The earliest philosophers were Greeks who occupied themselves mainly with the relations between God and Nature, and dealt with Man separately. The Christian Church was so absorbed in the relation of God to Man as entirely to neglect Nature. Finally, modern philosophers concern themselves chiefly with the relations between Man and Nature. Whether this is a correct historical generalization of the views which have been successively prevalent I do not care to discuss here, but the statement as to the scope of modern philosophy marks the limitations of Descartes' writings.

Descartes ended his book on geometry with the words: "I hope that posterity will judge me kindly, not only as to the things which I

LIVRE TROISIESME. 413

les Problesmes d'vn meſme genre, iay tout enſemble donné la façon de les reduire à vne infinité d'autres diuerſes; & ainſi de reſoudre chaſcun deux en vne infinité de façons. Puis outre cela qu'ayant conſtruit tous ceux qui ſont plans, en coupant d'vn cercle vne ligne droite; & tous ceux qui ſont ſolides, en coupant auſſy d'vn cercle vne Parabole; & enfin tous ceux qui ſont d'vn degré plus compoſés, en coupant tout de meſme d'vn cercle vne ligne qui n'eſt que d'vn degré plus compoſée que la Parabole; il ne faut que ſuiure la meſme voye pour conſtruire tous ceux qui ſont plus compoſés a l'infini. Car en matiere de progreſſions Mathematiques, lorſqu'on a les deux ou trois premiers termes, il n'eſt pas malayſé de trouuer les autres. Et i'eſpere que nos neueux me ſçauront gré, non ſeulement des choſes que iay icy expliquées; mais auſſy de celles que iay omiſes volontairerement, affin de leur laiſſer le plaiſir de les inuenter.

FIN.

Figure 26

have explained, but also as to those which I have intentionally omitted so as to leave to others the pleasure of discovery" (figure 26). He could not have known that a man who was only eight years old the year Descartes died would take his ideas of mathematics as the heart of science one huge step forward. This unsurpassed genius had more opportunities to experience the "pleasure of discovery" than probably any other individual in the history of the human race.

And There Was Light

The great eighteenth-century English poet Alexander Pope (1688–1744) was thirty-nine years old when Isaac Newton (1642–1727) died (figure 27 shows Newton's tomb inside Westminster Abbey).

In a well-known couplet, Pope attempted to encapsulate Newton's achievements:

Nature and Nature's laws lay hid in night:
God said, Let Newton be! And all was light.

Almost a hundred years after Newton's death, Lord Byron (1788–1824) added in his epic poem *Don Juan* the lines:

And this is the sole mortal who could grapple,
Since Adam, with a fall or with an apple.

To the generations of scientists that succeeded him, Newton indeed was and remains a figure of legendary proportions, even if one disregards the myths. Newton's famous quote "If I have seen further it is by standing on ye shoulders of Giants" is often presented as a model for the generosity and humility that scientists are expected to display about their greatest discoveries. Actually, Newton may have written that phrase as a subtly veiled sarcastic response to a letter from the person whom he regarded as his chief scientific nemesis, the prolific physicist and biologist Robert Hooke (1635–1703). Hooke had accused Newton

Figure 27

on several occasions of stealing his own ideas, first on the theory of light, and later on gravity. On January 20, 1676, Hooke adopted a more conciliatory tone, and in a personal letter to Newton he declared: "Your Designes and myne [concerning the theory of light] I suppose aim both at the same thing which is the Discovery of truth and I suppose we can both endure to hear objections." Newton decided to play the same game. In his reply to Hooke's letter, dated February 5, 1676, he wrote: "What Des-Cartes [Descartes] did was a good step [referring to Descartes' ideas on light]. You have added much several ways, & especially in taking the colours of thin plates into philosophical consideration. If I have seen further it is by standing on ye shoulders of Giants." Since, far from being a giant, Hooke was quite short and afflicted with a severe stoop, Newton's best-known quote might have simply meant that he felt he owed absolutely nothing to Hooke! The fact that Newton took every opportunity to insult Hooke, his statement that his own theory destroyed "all he [Hooke] has said," and his refusal to take his own book on light, *Opticks,* to the press until after Hooke's death, argue that this interpretation of the quote may not be too far-fetched. The feud between the two scientists reached an even higher peak when it came to the theory of gravity. When Newton heard that Hooke had claimed to be the originator of the law of gravity, he meticulously and vindictively erased every single reference to Hooke's name from the last part of his book on the subject. To his friend the astronomer Edmond Halley (1656–1742), Newton wrote on June 20, 1686:

> He [Hooke] should rather have excused himself by reason of his inability. For tis plain by his words he knew not how to go about it. Now is not this very fine? Mathematicians that find out, settle and do all the business must content themselves with being nothing but dry calculators and drudges and another that does nothing but pretend and grasp at all things must carry away all the invention as well of those that were to follow him as of those that went before.

Newton makes it abundantly clear here why he thought that Hooke did not deserve any credit—he could not formulate his ideas in the language of mathematics. Indeed, the quality that made Newton's theories truly stand out—the inherent characteristic that turned them into inevitable laws of nature—was precisely the fact that they were all expressed as crystal-clear, self-consistent mathematical relations. By comparison, Hooke's theoretical ideas, as ingenious as they were in many cases, looked like nothing but a collection of hunches, conjectures, and speculations.

Incidentally, the handwritten minutes of the Royal Society from 1661 to 1682, which were for a long time considered lost, suddenly surfaced in February 2006. The parchment, which contains more than 520 pages of script penned by Robert Hooke himself, was found in a house in Hampshire, England, where it is thought to have been stored in a cupboard for about fifty years. Minutes from December 1679 describe correspondence between Hooke and Newton in which they discussed an experiment to confirm the rotation of the Earth.

Returning to Newton's scientific masterstroke, Newton took Descartes' conception—that the cosmos can be described by mathematics—and turned it into a working reality. In the preface to his monumental work *The Mathematical Principles of Natural Philosophy* (in Latin: *Philosophiae Naturalis Principia Mathematica;* commonly known as *Principia*), he declared:

> We offer this work as the mathematical principles of philosophy, for the whole burden of philosophy seems to consist in this—from the phenomena of motions to investigate the forces of nature, and then from these forces to demonstrate the other phenomena: and to this end the general propositions in the first and second Books are directed. In the third Book we give an example of this in the explication of the System of the World; for by the propositions mathematically demonstrated in the former Books, in the third we derive from the celestial phenomena the force of gravity with which bodies tend to the Sun and the several planets. Then from these forces, by other propositions which are also mathematical, we deduce the motions of the planets, the comets, the moon, and the sea.

When we realize that Newton truly accomplished in *Principia* everything he promised in the preface, the only possible reaction is: Wow! Newton's innuendo of superiority to Descartes' work was also unmistakable: He chose the title of his book to read *Mathematical Principles*, as opposed to Descartes' *Principles of Philosophy*. Newton adopted the same mathematical reasoning and methodology even in his more experimentally based book on light, *Opticks*. He starts the book with: "My design in this book is not to explain the Properties of Light by Hypotheses, but to propose and prove them by Reason and Experiments: In order to which I shall premise the following definitions and Axioms." He then proceeds as if this were a book on Euclidean geometry, with concise definitions and propositions. Then, in the book's conclusion, Newton added for further emphasis: "As in Mathematicks, so in Natural Philosophy, the Investigation of difficult Things by the Method of Analysis, ought ever to precede the Method of Composition."

Newton's feat with his mathematical tool kit was nothing short of miraculous. This genius, who by a historical coincidence was born in exactly the same year in which Galileo died, formulated the fundamental laws of mechanics, deciphered the laws describing planetary motion, erected the theoretical basis for the phenomena of light and color, and founded the study of differential and integral calculus. These achievements alone would have sufficed to earn Newton a place of honor in the gallery of the most prominent scientists. But it was his work on gravity that elevated him to the top place on the podium of the magicians—the one reserved for the greatest scientist ever to have lived. That work bridged the gap between the heavens and the Earth, fused the fields of astronomy and physics, and put the entire cosmos under one mathematical umbrella. How was that masterpiece—*Principia*—born?

I Began to Think of Gravity Extending to the Orb of the Moon

William Stukeley (1687–1765), an antiquary and physician who was Newton's friend (in spite of the more than four decades in age separating them), eventually became the great scientist's first biographer. In his *Memoirs of Sir Isaac Newton's Life* we find an account of one of the most celebrated legends in the history of science:

> On 15 April 1726 I paid a visit to Sir Isaac at his lodgings in Orbels buildings in Kensington, dined with him and spent the whole day with him, alone . . . After dinner, the weather being warm, we went into the garden and drank thea, under the shade of some apple trees, only he and myself. Amidst other discourse, he told me he was just in the same situation, as when formerly [in 1666, when Newton returned home from Cambridge because of the plague], the notion of gravitation came into his mind. It was occasion'd by the fall of an apple, as he sat in contemplative mood. Why should that apple always descend perpendicularly to the ground, thought he to himself. Why should it not go sideways or upwards, but constantly to the earth's centre? Assuredly, the reason is, that the earth draws it. There must be a drawing power in matter: and the sum of the drawing power in the matter of the earth must be in the earth's centre, not in any side of earth. Therefore does this apple fall perpendicularly, or towards the centre. If matter thus draws matter, it must be in proportion of its quantity. Therefore the apple draws the earth, as well as the earth draws the apple. That there is a power, like that we here call gravity, which extends its self thro' the universe . . . This was the birth of those amazing discoverys, whereby he built philosophy on a solid foundation, to the astonishment of all Europe.

Irrespective of whether the mythical event with the apple actually occurred in 1666 or not, the legend sells Newton's genius and unique depth of analytic thinking rather short. While there is no doubt that Newton had written his first manuscript on the theory of gravity before 1669, he did not need to physically see a falling apple to know that the Earth attracted objects near its surface. Nor could his incredible insight in the formulation of a universal law of gravitation stem from the mere sight of a falling apple. In fact, there are some indications that a few crucial concepts that Newton needed to be able to enunciate a universally acting gravitational force were only conceived as late as 1684–85. An idea of this magnitude is so rare in the annals of science that even someone with a phenomenal mind—such as Newton—had to arrive at it through a long series of intellectual steps.

It may have all started in Newton's youth, with his less-than-perfect encounter with Euclid's massive treatise on geometry, *The Elements.* According to Newton's own testimony, he first "read only the titles of the propositions," since he found these so easy to understand that he "wondered how any body would amuse themselves to write any demonstrations of them." The first proposition that actually made him pause and caused him to introduce a few construction lines in the book was the one stating that "in a right triangle the square of the hypothenuse is equal to the squares of the two other sides"—the Pythagorean theorem. Somewhat surprisingly perhaps, even though Newton did read a few books on mathematics while at Trinity College in Cambridge, he did not read many of the works that were already available at this time. Evidently he didn't need to!

The one book that turned out to be perhaps the most influential in guiding Newton's mathematical and scientific thought was none other than Descartes' *La Géométrie*. Newton read it in 1664 and re-read it several times, until "by degrees he made himself master of the whole." The flexibility afforded by the notion of functions and their free variables appeared to open an infinitude of possibilities for Newton. Not only did analytic geometry pave the way for Newton's founding of calculus, with its associated exploration of functions, their tangents, and their curvatures, but Newton's inner scientific spirit was truly set ablaze. Gone were the dull constructions with ruler and compass—they were replaced by arbitrary curves that could be represented by algebraic expressions. Then in 1665–66, a horrible plague hit London. When the weekly death toll reached the thousands, the colleges of Cambridge had to close down. Newton was forced to leave school and return to his home in the remote village of Woolsthorpe. There, in the tranquility of the countryside, Newton made his first attempt to prove that the force that held the Moon in its orbit around the

Earth and the Earth's gravity (the very force that caused apples to fall) were, in fact, one and the same. Newton described those early endeavors in a memorandum written around 1714:

> And the same year [1666] I began to think of gravity extending to the orb of the Moon, and having found out how to estimate the force with which [a] globe revolving within a sphere presses the surface of the sphere, from Kepler's Rule of the periodical times of the Planets being in a sesquialternate proportion of their distances from the centres of their Orbs I deduced that the forces which keep the Planets in their Orbs must [be] reciprocally as the squares of their distances from the centres about which they revolve: and thereby compared the force requisite to keep the Moon in her Orb with the force of gravity at the surface of the earth, and found them answer pretty nearly. All this was in the two plague years of 1665 and 1666, for in those years I was in the prime of my age for invention, and minded Mathematicks and Philosophy more than at any time since.

Newton refers here to his important deduction (from Kepler's laws of planetary motion) that the gravitational attraction of two spherical bodies varies inversely as the square of the distance between them. In other words, if the distance between the Earth and the Moon were to be tripled, the gravitational force that the Moon would experience would be nine times (three squared) smaller.

For reasons that are not entirely clear, Newton essentially abandoned any serious research on the topics of gravitation and planetary motion until 1679. Then two letters from his archrival Robert Hooke renewed his interest in dynamics in general and in planetary motion in particular. The results of this revived curiosity were quite dramatic—using his previously formulated laws of mechanics, Newton proved Kepler's second law of planetary motion. Specifically, he showed that as the planet moves in its elliptical orbit about the Sun, the line joining the planet to the Sun sweeps equal areas in equal time intervals (figure 28). He also proved that for "a body revolving in an ellipse . . . the

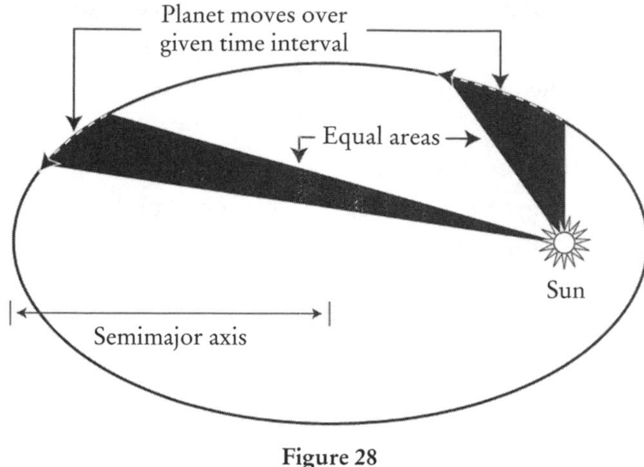

Figure 28

law of attraction directed to a focus of the ellipse . . . is inversely as the square of the distance." These were important milestones on the road to *Principia*.

Principia

Halley came to visit Newton in Cambridge in the spring or summer of 1684. For some time Halley had been discussing Kepler's laws of planetary motion with Hooke and with the renowned architect Christopher Wren (1632–1723). At these coffeehouse conversations, both Hooke and Wren claimed to have deduced the inverse-square law of gravity some years earlier, but both were also unable to construct a complete mathematical theory out of this deduction. Halley decided to ask Newton the crucial question: Did he know what would be the shape of the orbit of a planet acted upon by an attractive force varying as an inverse-square law? To his astonishment, Newton answered that he had proved some years earlier that the orbit would be an ellipse. The mathematician Abraham de Moivre (1667–1754) tells the story in a memorandum (from which a page is shown in figure 29):

> In 1684 Dr Halley came to visit him [Newton] at Cambridge, after they had been some time together, the D^r^ asked him what he thought the curve would be that would be described by the

> planets supposing the force of attraction towards the sun to be reciprocal to the square of their distance from it. Sr Isaac replied immediately that it would be an Ellipsis [ellipse], the Doctor struck with joy and amazement asked him how he knew it, why saith he [Newton] I have calculated it, whereupon Dr Halley asked him for his calculation without any farther delay, Sr Isaac looked among his papers but could not find it, but he promised him to renew it and send it.

Halley indeed came to visit Newton again in November 1684. Between the two visits Newton worked frantically. De Moivre gives us a brief description:

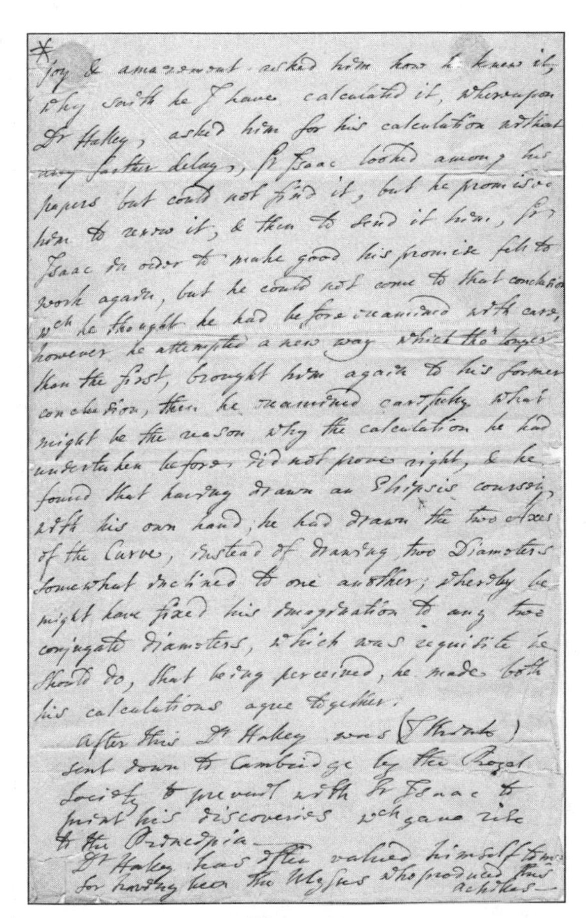
* joy & amazement asked him how he knew it,
why saith he I have calculated it, whereupon
Dr Halley asked him for his calculation without
any farther delay, Sr Isaac looked among his
papers but could not find it, but he promised
him to renew it, & then to send it him, Sr
Isaac in order to make good his promise fell to
work again, but he could not come to that conclusion
wch he thought he had before examined with care,
however he attempted a new way which tho' longer
than the first, brought him again to his former
conclusion, then he examined carefully what
might be the reason why the calculation he had
undertaken before did not prove right, & he
found that having drawn an Ellipsis coursely
with his own hand, he had drawn the two Axes
of the Curve, instead of drawing two Diameters
somewhat inclined to one another, whereby he
might have fixed his imagination to any two
conjugate diameters, which was requisite he
should do, that being perceived, he made both
his calculations agree together.
After this Dr Halley was (I think)
sent down to Cambridge by the Royal
Society to prevail with Sr Isaac to
print his discoveries wch gave rise
to the Principia.
Dr Halley has often valued himself [illegible]
for having been the Ulysses who produced this
Achilles.

Figure 29

> Sr Isaac in order to make good his promise fell to work again but he could not come to that conclusion wch he thought he had before examined with care, however he attempted a new way which tho^{u} longer than the first, brought him again to his former conclusion, then he examined carefully what might be the reason why the calculation he had undertaken before did not prove right, & . . . he made both his calculations agree together.

This dry summary does not even begin to tell us what Newton had actually accomplished in the few months between Halley's two visits. He wrote an entire treatise, *De Motu Corporum in Gyrum* (*The Motion of Revolving Bodies*), in which he proved most aspects of bodies moving in circular or elliptical orbits, proved all of Kepler's laws, and even solved for the motion of a particle moving in a resisting medium (such as air). Halley was overwhelmed. To his satisfaction, he at least managed to persuade Newton to publish all of these staggering discoveries—*Principia* was finally about to happen.

At first, Newton had thought of the book as being nothing but a somewhat expanded and more detailed version of his treatise *De Motu*. As he started working, however, he realized that some topics required further thought. Two points in particular continued to disturb Newton. One was the following: Newton originally formulated his law of gravitational attraction as if the Sun, Earth, and planets were mathematical point masses, without any dimensions. He of course knew this not to be true, and therefore he regarded his results as only approximate when applied to the solar system. Some even speculate that he abandoned again his pursuit of the topic of gravity after 1679 because of his dissatisfaction with this state of affairs. The situation was even worse with respect to the force on the apple. There, clearly the parts of the Earth that are directly underneath the apple are at a much shorter distance to it than the parts that are on the other side of the Earth. How was one to calculate the net attraction? The astronomer Herbert Hall Turner (1861–1930) described Newton's mental struggle in an article that appeared in the London *Times* on March 19, 1927:

> At that time the general idea of an attraction varying as the inverse square of the distance occurred to him, but he saw grave difficulties in its complete application of which lesser minds were unconscious. The most important of these he did not overcome until 1685 . . . It was that of linking up the attraction of the earth on a body so far away as the moon with its attraction on the apple close to its surface. In the former case the various particles composing the earth (to which individually Newton hoped to extend his law, thus making it universal) are at distances from the moon not greatly different either in magnitude or direction; but their distances from the apple differ conspicuously in both size and direction. How are the separate attractions in the latter case to be added together or combined into a single resultant? And in what "centre of gravity," if any, may they be concentrated?

The breakthrough finally came in the spring of 1685. Newton managed to prove an essential theorem: For two spherical bodies, "the whole force with which one of these spheres attracts the other will be inversely proportional to the square of the distance of the centres." That is, spherical bodies gravitationally act as if they were point masses concentrated at their centers. The importance of this beautiful proof was emphasized by the mathematician James Whitbread Lee Glaisher (1848–1928). In his address at the bicentenary celebration (in 1887) of Newton's *Principia*, Glaisher said:

> No sooner had Newton proved this superb theorem—and we know from his own words that he had no expectation of so beautiful a result till it emerged from his mathematical investigation—than all the mechanism of the universe at once lay spread before him . . . How different must these propositions have seemed to Newton's eyes when he realised that these results, which he had believed to be only approximately true when applied to the solar system, were really exact! . . . We can imagine the effect of this sudden transition from approximation to exactitude in stimulating Newton's mind to still greater

> efforts. It was now in his power to apply mathematical analysis with absolute precision to the actual problem of astronomy.

The other point that was apparently still troubling Newton when he wrote the early draft of *De Motu* was the fact that he neglected the influence of the forces by which the planets attracted the Sun. In other words, in his original formulation, he reduced the Sun to a mere unmovable center of force of the type that "hardly exists," in Newton's words, in the real world. This scheme contradicted Newton's own third law of motion, according to which "the actions of attracting and attracted bodies are always mutual and equal." Each planet attracts the Sun precisely with the same force that the Sun attracts the planet. Consequently, he added, "if there are two bodies [such as the Earth and the Sun], neither the attracting nor the attracted body can be at rest." This seemingly minor realization was actually an important stepping-stone toward the concept of a universal gravity. We can attempt to guess Newton's line of thought: If the Sun pulls the Earth, then the Earth must also pull the Sun, with equal strength. That is, the Earth doesn't simply orbit the Sun, but rather they both revolve around their mutual center of gravity. But this is not all. All the other planets also attract the Sun, and indeed each planet feels the attraction not just of the Sun, but also of all other planets. The same type of logic could be applied to Jupiter and its satellites, to the Earth and the Moon, and even to an apple and the Earth. The conclusion is astounding in it simplicity—*there is only one gravitational force, and it acts between any two masses, anywhere in the universe*. This was all that Newton needed. The *Principia*—510 dense Latin pages—was published in July of 1687.

Newton took observations and experiments that were accurate to only about 4 percent and established from those a mathematical law of gravity that turned out to be accurate to better than one part in a million. He united for the first time *explanations* of natural phenomena with the power of *prediction* of the results of observations. Physics and mathematics became forever intertwined, while the divorce of science from philosophy became inevitable.

The second edition of the *Principia,* edited extensively by Newton and in particular by the mathematician Roger Cotes (1682–1716), appeared in 1713 (figure 30 shows the frontispiece). Newton, who was never known for his warmth, did not even bother to thank Cotes in the preface to the book for his fabulous work. Still, when Cotes died from violent fever at age thirty-three, Newton did show some appreciation: "If he had lived we would have known something."

Curiously, some of Newton's most memorable remarks about God appeared only as afterthoughts in the second edition. In a letter to Cotes on March 28, 1713, less than three months before the completion of *Principia*'s second edition, Newton included the sentence: "It surely does belong to natural philosophy to discourse of God from the phenomena [of nature]." Indeed, Newton expressed his ideas of an "eternal and infinite, omnipotent and omniscient" God in the "General Scholium"—the section he regarded as putting the final touch on the *Principia.*

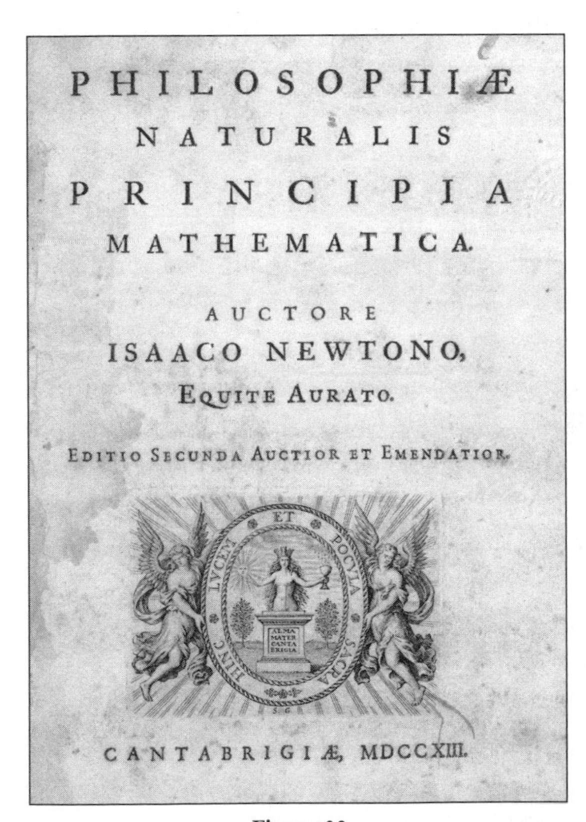

PHILOSOPHIÆ
NATURALIS
PRINCIPIA
MATHEMATICA.

AUCTORE
ISAACO NEWTONO,
EQUITE AURATO.

EDITIO SECUNDA AUCTIOR ET EMENDATIOR.

CANTABRIGIÆ, MDCCXIII.

Figure 30

But did God's role remain unchanged in this increasingly mathematical universe? Or was God perceived more and more as a mathematician? After all, until the formulation of the law of gravitation, the motions of the planets had been regarded as one of the unmistakable works of God. How did Newton and Descartes see this shift in emphasis toward scientific explanations of nature?

The Mathematician God of Newton and Descartes

As were most people of their time, both Newton and Descartes were religious men. The French writer known by the pen name of Voltaire (1694–1778), who wrote quite extensively about Newton, famously said that "if God did not exist, it would be necessary for us to invent Him."

For Newton, the world's very existence and the mathematical regularity of the observed cosmos were evidence for God's presence. This type of causal reasoning was first used by the theologian Thomas Aquinas (ca. 1225–1274), and the arguments fall under the general philosophical labels of a *cosmological argument* and a *teleological argument.* Put simply, the cosmological argument claims that since the physical world had to come into existence somehow, there must be a First Cause, namely, a creator God. The teleological argument, or *argument from design,* attempts to furnish evidence for God's existence from the apparent design of the world. Here are Newton's thoughts, as expressed in *Principia:* "This most beautiful system of the sun, planets and comets, could only proceed from the counsel and dominion of an intelligent and powerful Being. And if the fixed stars are the centers of other like systems, these, being formed by the like wise counsel, must be all subject to the dominion of One." The validity of the cosmological, teleological, and similar arguments as proof for God's existence has been the subject of debate among philosophers for centuries. My personal impression has always been that theists don't need these arguments to be convinced, and atheists are not persuaded by them.

Newton added yet another twist, based on the universality of his laws. He regarded the fact that the entire cosmos is governed by the same laws and appears to be stable as further evidence for God's guiding hand, "especially since the light of the fixed stars is of the *same nature* [emphasis added] with the light of the sun, and from every system light passes into all the other systems: and lest the systems of the fixed stars should, by their gravity, fall on each other mutually, he hath placed these systems at immense distances one from another."

In his book *Opticks,* Newton made it clear that he did not believe that the laws of nature by themselves were sufficient to explain the universe's existence—God was the creator and sustainer of all the atoms that make up the cosmic matter: "For it became him [God] who created them [the atoms] to set them in order. And if he did so, it's unphilosophical to seek for any other Origin of the World, or to pretend that it might arise out of a Chaos by the mere Laws of Nature." In other words, to Newton, God was a mathematician (among other things), not just as a figure of speech, but almost literally—the Creator God brought into existence a physical world that was governed by mathematical laws.

Being much more philosophically inclined than Newton, Descartes had been extremely preoccupied with proving God's existence. To him, the road from the certainty in our own existence ("I am thinking, therefore I exist") to our ability to construct a tapestry of objective science had to pass through an unassailable proof for the existence of a supremely perfect God. This God, he argued, was the ultimate source of all truth and the only guarantor of the reliability of human reasoning. This suspiciously circular argument (known as the *Cartesian circle*) was already criticized during Descartes' time, especially by the French philosopher, theologian, and mathematician Antoine Arnauld (1612–94). Arnauld posed a question that was devastating in its simplicity: If we need to prove God's existence in order to guarantee the validity of the human thought process, how can we trust the proof, which is in itself a product of the human mind? While Descartes did make some desperate attempts to escape from this vicious reasoning circle, many of the philosophers who followed him did not find his efforts particularly convincing. Descartes' "supplemental proof" for the existence of God was equally questionable. It falls under the general philosophical label of an *ontological argument.*

The philosophical theologian St. Anselm of Canterbury (1033–1109) first formulated this type of reasoning in 1078, and it has since resurfaced in many incarnations. The logical construct goes something like this: God, by definition, is so perfect that he is the greatest conceivable being. But if God did not exist, then it would be possible to conceive of a greater being yet—one that in addition to being blessed with all of God's perfections also exists. This would contradict God's definition as the greatest conceivable being—therefore God has to exist. In Descartes' words: "Existence can no more be separated from the essence of God than the fact that its angles equal two right angles can be separated from the essence of a triangle."

This type of logical maneuvering does not convince many philosophers, and they argue that to establish the existence of anything that is consequential in the physical world, and in particular something as grand as God, logic alone does not suffice.

Oddly enough, Descartes was accused of fostering atheism, and his works were put on the Catholic Church's Index of Forbidden Books in 1667. This was a bizarre charge in light of Descartes' insistence on God as the ultimate guarantor of truth.

Leaving the purely philosophical questions aside, for our present purposes the most interesting point is Descartes' view that God created all the "eternal truths." In particular, he declared that "the mathematical truths which you call eternal have been laid down by God and depend on Him entirely no less than the rest of his creatures." So the Cartesian God was more than a mathematician, in the sense of being the creator of both mathematics and a physical world that is entirely based on mathematics. According to this worldview, which was becoming prevalent at the end of the seventeenth century, humans clearly only *discover* mathematics and do not invent it.

More significantly, the works of Galileo, Descartes, and Newton have changed the relationship between mathematics and the sciences in a profound way. First, the explosive developments in science became strong motivators for mathematical investigations. Second, through Newton's laws, even more abstract mathematical fields, such as calculus, became the *essence* of physical explanations. Finally, and perhaps most importantly, the boundary between mathematics and the sciences was blurred beyond recognition, almost to the point of a complete fusion between mathematical insights and large swaths of exploration. All of these developments created a level of enthusiasm for mathematics perhaps not experienced since the time of the ancient Greeks. Mathematicians felt that the world was theirs to conquer, and that it offered unlimited potential for discovery.

CHAPTER 5

STATISTICIANS AND PROBABILISTS: THE SCIENCE OF UNCERTAINTY

The world doesn't stand still. Most things around us are either in motion or continuously changing. Even the seemingly firm Earth underneath our feet is in fact spinning around its axis, revolving around the Sun, and traveling (together with the Sun) around the center of our Milky Way galaxy. The air we breathe is composed of trillions of molecules that move ceaselessly and randomly. At the same time, plants grow, radioactive materials decay, the atmospheric temperature rises and falls both daily and with the seasons, and the human life expectancy keeps increasing. This cosmic restlessness in itself, however, did not stump mathematics. The branch of mathematics called *calculus* was introduced by Newton and Leibniz precisely to permit a rigorous analysis and an accurate modeling of both motion and change. By now, this incredible tool has become so potent and all encompassing that it can be used to examine problems as diverse as the motion of the space shuttle or the spreading of an infectious disease. Just as a movie can capture motion by breaking it up into a frame-by-frame sequence, calculus can measure change on such a fine grid that it allows for the determination of quantities that have only a fleeting existence, such as instantaneous speed, acceleration, or rate of change.

Continuing in Newton's and Leibniz's giant footsteps, mathematicians of the Age of Reason (the late seventeenth and eighteenth centuries) extended calculus to the even more powerful and widely applicable branch of *differential equations*. Armed with this new weapon, scientists were now able to present detailed mathematical theories of phenomena ranging from the music produced by a violin string to the transport of heat, from the motion of a spinning top to the flow of liquids and gases. For a while, differential equations became the tool of choice for making progress in physics.

A few of the first explorers of the new vistas opened by differential equations were members of the legendary Bernoulli family. Between the mid-seventeenth century and the mid-nineteenth century, this family produced no fewer than eight prominent mathematicians. These gifted individuals were almost equally known for their bitter intrafamily feuds as they were for their outstanding mathematics. While the Bernoulli quarrels were always concerned with competition for mathematical supremacy, some of the problems they argued about may not seem today to be of the highest significance. Still, the solution of these intricate puzzles often paved the way for more impressive mathematical breakthroughs. Overall, there is no question that the Bernoullis played an important role in establishing mathematics as the language of a variety of physical processes.

One story can help exemplify the complexity of the minds of two of the brightest Bernoullis—the brothers Jakob (1654–1705) and Johann (1667–1748). Jakob Bernoulli was one of the pioneers of *probability theory*, and we shall return to him later in the chapter. In 1690, however, Jakob was busy resurrecting a problem first examined by the quintessential Renaissance man, Leonardo da Vinci, two centuries earlier: What is the shape taken by an elastic but inextensible chain suspended from two fixed points (as in figure 31)? Leonardo sketched a few such chains in his notebooks. The problem was also presented to Descartes by his friend Isaac Beeckman, but there is no evidence of Descartes' trying to solve it. Eventually the problem became known as the problem of the *catenary* (from the Latin word *catena*, meaning "a chain"). Galileo thought that the shape would be parabolic but was proven wrong by the French Jesuit Ignatius Pardies (1636–73). Pardies

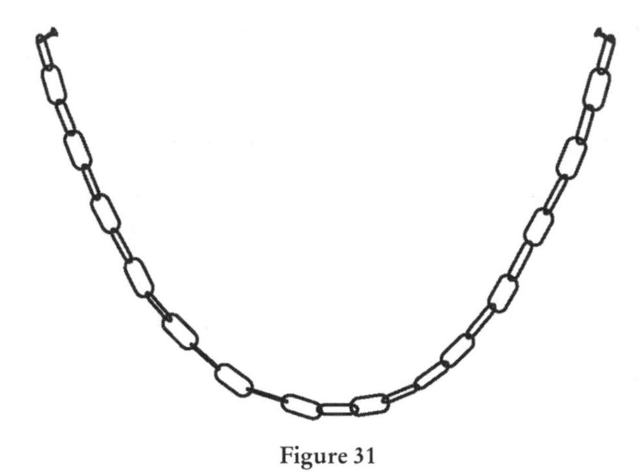

Figure 31

was not up to the task, however, of actually solving mathematically for the correct shape.

Just one year after Jakob Bernoulli posed the problem, his younger brother Johann solved it (by means of a differential equation). Leibniz and the Dutch mathematical physicist Christiaan Huygens (1629–95) also solved it, but Huygens's solution employed a more obscure geometrical method. The fact that Johann managed to solve a problem that had stymied his brother and teacher continued to be an immense source of satisfaction to the younger Bernoulli, even as late as thirteen years after Jakob's death. In a letter Johann wrote on September 29, 1718, to the French mathematician Pierre Rémond de Montmort (1678–1719), he could not hide his delight:

> You say that my brother proposed this problem; that is true, but does it follow that he had a solution of it then? Not at all. When he proposed this problem at my suggestion (for I was the first to think of it), neither the one nor the other of us was able to solve it; we despaired of it as insoluble, until Mr. Leibniz gave notice to the public in the Leipzig journal of 1690, p. 360, that he had solved the problem but did not publish his solution, so as to give time to other analysts, and it was this that encouraged us, my brother and me, to apply ourselves afresh.

After shamelessly taking ownership of even the suggestion of the problem, Johann continued with unconcealed glee:

> The efforts of my brother were without success; for my part, I was more fortunate, for I found the skill (I say it without boasting, why should I conceal the truth?) to solve it in full . . . It is true that it cost me study that robbed me of rest for an entire night . . . but the next morning, filled with joy, I ran to my brother, who was still struggling miserably with this Gordian knot without getting anywhere, always thinking like Galileo that the catenary was a parabola. Stop! Stop! I say to him, don't torture yourself any more to try to prove the identity of the catenary with the parabola, since it is entirely false . . . But then you astonish me by concluding that my brother found a method of solving this problem . . . I ask you, do you really think, if my brother had solved the problem in question, he would have been so obliging to me as not to appear among the solvers, just so as to cede me the glory of appearing alone on the stage in the quality of the first solver, along with Messrs. Huygens and Leibniz?

In case you ever needed proof that mathematicians are humans after all, this story amply provides it. The familial rivalry, however, does not take anything away from the accomplishments of the Bernoullis. During the years that followed the catenary episode, Jakob, Johann, and Daniel Bernoulli (1700–1782) went on not only to solve other similar problems of hanging cords, but also to advance the theory of differential equations in general and to solve for the motion of projectiles through a resisting medium.

The tale of the catenary serves to demonstrate another facet of the power of mathematics—even seemingly trivial physical problems have mathematical solutions. The shape of the catenary itself, by the way, continues to delight millions of visitors to the famous Gateway Arch in St. Louis, Missouri. The Finnish-American architect Eero Saarinen (1910–61) and the German-American structural engineer

Hannskarl Bandel (1925–93) designed this iconic structure in a shape that is similar to that of an inverted catenary.

The astounding success of the physical sciences in discovering mathematical laws that govern the behavior of the cosmos at large raised the inevitable question of whether or not similar principles might also underlie biological, social, or economical processes. Is mathematics only the language of nature, mathematicians wondered, or is it also the language of human nature? Even if truly universal principles do not exist, could mathematical tools, at the very least, be used to model and subsequently explain social behavior? At first, many mathematicians were quite convinced that "laws" based on some version of calculus would be able to accurately predict all future events, large or small. This was the opinion, for instance, of the great mathematical physicist Pierre-Simon de Laplace (1749–1827). Laplace's five volumes of *Mécanique céleste* (*Celestial Mechanics*) gave the first virtually complete (if approximate) solution to the motions in the solar system. In addition, Laplace was the man who answered a question that puzzled even the giant Newton: Why is the solar system as stable as it is? Newton thought that due to their mutual attractions planets had to fall into the Sun or to fly away into free space, and he invoked God's hand in keeping the solar system intact. Laplace had rather different views. Instead of relying on God's handiwork, he simply proved mathematically that the solar system is stable over periods of time that are much longer than those anticipated by Newton. To solve this complex problem, Laplace introduced yet another mathematical formalism known as *perturbation theory,* which enabled him to calculate the cumulative effect of many small perturbations to each planet's orbit. Finally, to top it all, Laplace proposed one of the first models for the very origin of the solar system—in his influential *nebular hypothesis*, the solar system formed from a contracting gaseous nebula.

Given all of these impressive feats, it is perhaps not surprising that in his *Philosophical Essay on Probabilities* Laplace boldly pronounced:

> All events, even those which on account of their insignificance do not seem to follow the great laws of nature, are a result of it just as necessary as the revolutions of the Sun. In ignorance

> of the ties which unite such events to the entire system of the universe, they have been made to depend upon final causes for or upon hazard . . . We ought then to regard the present state of the universe as the effect of its anterior state and as the cause of the one which is to follow. Given for one instant an intelligence which could comprehend all the forces by which nature is animated and the respective situations of the beings who compose it—an intelligence sufficiently vast to submit these data to analysis—it would embrace in the same formula the movements of the greatest bodies of the universe and those of the lightest atom; for it, nothing would be uncertain and the future, as the past, would be present to its eyes. The human mind offers, in the perfection which it has been able to give to astronomy, a feeble idea of this intelligence.

Just in case you wonder, when Laplace talked about this hypothetical supreme "intelligence," he did not mean God. Unlike Newton and Descartes, Laplace was not a religious person. When he gave a copy of his *Celestial Mechanics* to Napoleon Bonaparte, the latter, who had heard that there was no reference to God in the work, remarked: "M. Laplace, they tell me you have written this huge book on the system of the universe and have never even mentioned its creator." Laplace immediately replied: "I did not need to make that hypothesis." The amused Napoleon told the mathematician Joseph-Louis Lagrange about this reply, and the latter exclaimed: "Ah! That is a beautiful hypothesis; it explains many things." But the story doesn't end there. When he heard about Lagrange's reaction, Laplace commented dryly: "This hypothesis, Sir, explains in fact everything, but does not permit to predict anything. As a scholar, I must provide you with works permitting predictions."

The twentieth century development of quantum mechanics—the theory of the subatomic world—has proven the expectation for a fully deterministic universe to be too optimistic. Modern physics has in fact demonstrated that it is impossible to predict the outcome of every experiment, even in principle. Rather, the theory can only predict the probabilities for different results. The situation in the social sciences is clearly even more complex because of a multiplicity of interrelated elements, many of which are highly uncertain at best. The researchers of the seventeenth century realized soon enough that a search for precise universal social principles of the type of Newton's law of gravitation was doomed from the start. For a while, it seemed that when the intricacies of human nature are brought into the equation, secure predictions become virtually impossible. The situation appeared to be even more hopeless when the minds of an entire population were involved. Rather than despairing, however, a few ingenious thinkers developed a fresh arsenal of innovative mathematical tools—*statistics* and *probability theory.*

The Odds Beyond Death and Taxes

The English novelist Daniel Defoe (1660–1731), best known for his adventure story *Robinson Crusoe,* also authored a work on the supernatural entitled *The Political History of the Devil.* In it, Defoe, who saw evidence for the devil's actions everywhere, wrote: "Things as certain as death and taxes, can be more firmly believed." Benjamin Franklin (1706–90) seems to have subscribed to the same perspective with respect to certainty. In a letter he wrote at age eighty-three to the French physicist Jean-Baptiste Leroy, he said: "Our Constitution is in actual operation. Everything appears to promise that it will last; but in this world nothing can be said to be certain but death and taxes." Indeed, the courses of our lives appear to be unpredictable, prone to natural disasters, susceptible to human errors, and affected by pure happenstance. Phrases such as "[- - - -] happens" have been invented precisely to express our vulnerability to the unexpected and our inability to control chance. In spite of these obstacles, and maybe even because of these challenges, mathematicians, social scientists, and biologists have embarked since the sixteenth century on serious attempts to tackle uncertainties methodically. Following the establishment of the field of statistical mechanics, and faced with the realization that the very foundations of physics—in the form of quantum mechanics—are based on uncertainty, physicists of the twentieth and twenty-first centuries have enthusiastically joined the battle. The weapon researchers use to combat the lack of precise determinism is the ability to calculate the odds of a particular outcome. Short of being capable of actually predicting a result, computing the likelihood of different consequences is the next best thing. The tools that have been fashioned to improve on mere guesses and speculations—statistics and probability theory—provide the underpinning of not just much of modern science, but also a wide range of social activities, from economics to sports.

We all use probabilities and statistics in almost every decision we make, sometimes subconsciously. For instance, you probably don't know that the number of fatalities from automobile accidents in the U.S. was 42,636 in 2004. However, had that number been, say, 3 million, I'm sure you would have known about it. Furthermore, this knowledge would have probably caused you to think twice before getting into the car in the morning. Why do these precise data on road fatalities give us some confidence in our decision to drive? As we shall see shortly, a key ingredient to their reliability is the fact that they are based on very large numbers. The number of fatalities in Frio Town, Texas, with a population of forty-nine in 1969 would hardly have been equally convincing. Probability and statistics are among the most important arrows for the bows of economists, political consultants, geneticists, insurance companies, and anybody trying to distill meaningful conclusions from vast amounts of data. When we talk about mathematics permeating even disciplines that were not originally under the umbrella of the exact sciences, it is often through the windows opened by probability theory and statistics. How did these fruitful fields emerge?

Statistics—a term derived from the Italian *stato* (state) and *statista* (a person dealing with state affairs)—first referred to the simple collection of facts by government officials. The first important work on statistics in the modern sense was carried out by an unlikely researcher—a shopkeeper in seventeenth century London. John Graunt (1620–74) was trained to sell buttons, needles, and drapes. Since his job afforded him a considerable amount of free time, Graunt studied Latin and French on his own and started to take interest in the Bills of Mortality—weekly numbers of deaths parish by parish—that had been published in London since 1604. The process of issuing these reports was established mainly in order to provide an early warning signal for devastating epidemics. Using those crude numbers, Graunt started to make interesting observations that he eventually published in a small, eighty-five-page book entitled *Natural and Political Observations Mentioned in a Following Index, and Made upon the Bills of Mortality.* Figure 32 presents an example of a table from Graunt's book, where no fewer than sixty-three diseases and casualties were listed alphabetically. In a dedication to the president of the Royal Society, Graunt points out that since his work concerns "the Air, Countries, Seasons, Fruitfulness, Health, Diseases, Longevity, and the proportion between the Sex and Ages of Mankind," it is really a treatise in

(9)

The Diſeaſes, and Caſualties this year being 1632.

Abortive, and Stilborn	445	Jaundies	43
Affrighted	1	Jawfaln	8
Aged	628	Impoſtume	74
Ague	43	Kil'd by ſeveral accidents	46
Apoplex, and Meagrom	17	King's Evil	38
Bit with a mad dog	1	Lethargie	2
Bleeding	3	Livergrown	87
Bloody flux, ſcowring, and flux	348	Lunatique	5
Bruſed, Iſſues, ſores, and ulcers,	28	Made away themſelves	15
Burnt, and Scalded	5	Meaſles	80
Burſt, and Rupture	9	Murthered	7
Cancer, and Wolf	10	Over-laid, and ſtarved at nurſe	7
Canker	1	Palſie	25
Childbed	171	Piles	1
Chriſomes, and Infants	2268	Plague	8
Cold, and Cough	55	Planet	13
Colick, Stone, and Strangury	56	Pleuriſie, and Spleen	36
Conſumption	1797	Purples, and ſpotted Feaver	38
Convulſion	241	Quinſie	7
Cut of the Stone	5	Riſing of the Lights	98
Dead in the ſtreet, and ſtarved	6	Sciatica	1
Dropſie, and Swelling	267	Scurvey, and Itch	9
Drowned	34	Suddenly	62
Executed, and preſt to death	18	Surfet	86
Falling Sickneſs	7	Swine Pox	6
Fever	1108	Teeth	470
Fiſtula	13	Thruſh, and Sore mouth	40
Flocks, and ſmall Pox	531	Tympany	13
French Pox	12	Tiſſick	34
Gangrene	5	Vomiting	1
Gout	4	Worms	27
Grief	11		

Chriſtened { Males—4994, Females-4590, In all—9584 } Buried { Males —4932, Females—4603, In all —9535 } Whereof, of the Plague-8

Increaſed in the Burials in the 122 Pariſhes, and at the Peſthouſe this year 993
Decreaſed of the Plague in the 122 Pariſhes, and at the Peſthouſe this year, 266

C 7 In

Figure 32

natural history. Indeed, Graunt did much more than merely collect and present the data. By examining, for instance, the average numbers of christenings and burials for males and females in London and in the country parish Romsey in Hampshire, he demonstrated for the first time the stability of the sex ratio at birth. Specifically, he found that in London there were thirteen females born for every fourteen males and in Romsey fifteen females for sixteen males. Remarkably, Graunt had the foresight to express the wish that "travellers would enquire whether it be the same in other countries." He also noted that "it is a blessing to Man-kind, that by this overplus of *Males* there is this natural Bar to *Polygamy*: for in such a state Women could not live in that parity, and equality of expence with their Husbands, as now, and here they do." Today, the commonly assumed ratio between boys and girls at birth is about 1.05. Traditionally the explanation for this excess of males is that Mother Nature stacks the deck in favor of male births because of the somewhat greater fragility of male fetuses and babies. Incidentally, for reasons that are not entirely clear, in both the United States and Japan the proportion of baby boys has fallen slightly each year since the 1970s.

Another pioneering effort by Graunt was his attempt to construct an age distribution, or a "life table," for the living population, using the data on the number of deaths according to cause. This was clearly of great political importance, since it had implications for the number of fighting men—men between sixteen and fifty-six years of age—in the population. Strictly speaking, Graunt did not have sufficient information to deduce the age distribution. This is precisely where, however, he demonstrated ingenuity and creative thinking. Here is how he describes his estimate of childhood mortality:

> Our first Observation upon the Casualties shall be, that in twenty Years there dying of all diseases and Casualties, 229,250, that 71,124 dyed of the Thrush, Convulsion, Rickets, Teeths, and Worms; and as Abortives, Chrysomes, Infants, Livergrown, and Overlaid; that is to say, that about ⅓ of the whole died of those diseases, which we guess did all light upon Children under four or five Years old. There died also of the Small-Pox,

> Swine-Pox, and Measles, and of Worms without Convulsions, 12,210, of which number we suppose likewise that about ½ might be Children under six Years old. Now, if we consider that 16 of the said 229 thousand died of that extraordinary and grand Casualty the Plague, we shall finde that about thirty six percentum of all quick conceptions, died before six years old."

In other words, Graunt estimated the mortality before age six to be (71,124 + 6,105) ÷ (229,250 – 16,000) = 0.36. Using similar arguments and educated guesses, Graunt was able to estimate the old-age mortality. Finally, he filled the gap between ages six and seventy-six by a mathematical assumption about the behavior of the mortality rate with age. While many of Graunt's conclusions were not particularly sound, his study launched the science of statistics as we know it. His observation that the percentages of certain events previously considered purely a matter of chance or fate (such as deaths caused by various diseases) in fact showed an extremely robust regularity, introduced scientific, quantitative thinking into the social sciences.

The researchers who followed Graunt adopted some aspects of his methodology, but also developed a better mathematical understanding of the use of statistics. Surprisingly perhaps, the person who made the most significant improvements to Graunt's life table was the astronomer Edmond Halley—the same person who persuaded Newton to publish his *Principia.* Why was everybody so interested in life tables? Partly because this was, and still is, the basis for life insurance. Life insurance companies (and indeed gold diggers who marry for money!) are interested in such questions as: If a person lived to be sixty, what is the probability that he or she would also live to be eighty?

To construct his life table, Halley used detailed records that were kept at the city of Breslau in Silesia since the end of the sixteenth century. A local pastor in Breslau, Dr. Caspar Neumann, was using those lists to suppress superstitions in his parish that health is affected by the phases of the Moon or by ages that are divisible by seven and nine. Eventually, Halley's paper, which had the rather long title of "An Estimate of the Degrees of the Mortality of Mankind, drawn from curious Tables of the Births and Funerals at the City of Breslaw; with an Attempt to ascertain the Price of Annuities upon Lives," became the basis for the mathematics of life insurance. To get an idea of how insurance companies may assess their odds, examine Halley's life table below:

Halley's Life Table

AGE CURRENT	PERSONS	AGE CURRENT	PERSONS	AGE CURRENT	PERSONS
1	1000	11	653	21	592
2	855	12	646	22	586
3	798	13	640	23	579
4	760	14	634	24	573
5	732	15	628	25	567
6	710	16	622	26	560
7	692	17	616	27	553
8	680	18	610	28	546
9	670	19	604	29	539
10	661	20	598	30	531
AGE CURRENT	PERSONS	AGE CURRENT	PERSONS	AGE CURRENT	PERSONS
31	523	41	436	51	335
32	515	42	427	52	324
33	507	43	417	53	313
34	499	44	407	54	302
35	490	45	397	55	292
36	481	46	387	56	282
37	472	47	377	57	272
38	463	48	367	58	262
39	454	49	357	59	252
40	445	50	346	60	242
AGE CURRENT	PERSONS	AGE CURRENT	PERSONS	AGE CURRENT	PERSONS
61	232	71	131	81	34
62	222	72	120	82	28
63	212	73	109	83	23
64	202	74	98	84	20
65	192	75	88		
66	182	76	78		
67	172	77	68		
68	162	78	58		
69	152	79	49		
70	142	80	41		

The table shows, for instance, that of 710 people alive at age six, 346 were still alive at age fifty. One could then take the ratio of 346/710 or 0.49 as an estimate of the probability that a person of age six would live to be fifty. Similarly, of 242 at age sixty, 41 were alive at age eighty. The probability of making it from sixty to eighty could then be estimated to be 41/242, or about 0.17. The rationale behind this procedure is simple. It relies on past experience to determine the probability of various future events. If the sample on which the experience is predicated is sufficiently large (Halley's table was based on a population of about 34,000), and if certain assumptions hold (such as that the mortality rate is constant over time), then the calculated probabilities are fairly reliable. Here is how Jakob Bernoulli described the same problem:

> What mortal, I ask, could ascertain the number of diseases, counting all possible cases, that afflict the human body in every one of its many parts and at every age, and say how much more likely one disease is to be fatal than another . . . and on that basis make a prediction about the relationship between life and death in future generations?

After concluding that this and similar forecasts "depend on factors that are completely obscure, and which constantly deceive our senses by the endless complexity of their interrelationships," Bernoulli also suggested a statistical/probabilistic approach:

> There is, however, another way that will lead us to what we are looking for and enable us at least to ascertain *a posteriori* what we cannot determine *a priori,* that is, to ascertain it from the results observed in numerous similar instances. It must be assumed in this connection that, under similar conditions, the occurrence (or nonoccurrence) of an event in the future will follow the same pattern as was observed for like events in the past. For example, if we have observed that out of 300 persons of the same age and with the same constitution as a certain *Titius,* 200 died within ten years while the rest survived, we can with reasonable certainty conclude that there are twice

> as many chances that Titius also will have to pay his debt to nature within the ensuing decade as there are chances that he will live beyond that time.

Halley followed his mathematical articles on mortality with an interesting note that had more philosophical overtones. One of the passages is particularly moving:

> Besides the uses mentioned in my former, it may perhaps not be an unacceptable thing to infer from the same Tables, how unjustly we repine at the shortness of our lives, and think our selves wronged if we attain not Old Age; whereas it appears hereby, that the one half of those that are born are dead in Seventeen years time, 1238 being in that time reduced to 616. So that instead of murmuring at what we call an untimely Death, we ought with Patience and unconcern to submit to that Dissolution which is the necessary Condition of our perishable Materials, and of our nice and frail Structure and Composition: And to account it as Blessing that we have survived, perhaps by many Years, that Period of Life, whereat the one half of the whole Race of Mankind does not arrive.

While the situation in much of the modern world has improved significantly compared to Halley's sad statistics, this is unfortunately not true for all countries. In Zambia, for instance, the mortality for ages five and under in 2006 has been estimated at a staggering 182 deaths per 1,000 live births. The life expectancy in Zambia remains at a heartbreaking low of thirty-seven years.

Statistics, however, are not concerned only with death. They penetrate into every aspect of human life, from mere physical traits to intellectual products. One of the first to recognize the power of statistics to potentially produce "laws" for the social sciences was the Belgian polymath Lambert-Adolphe-Jacques Quetelet (1796–1874). He, more than anyone else, was responsible for the introduction of the common statistical concept of the "average man," or what we would refer to today as the "average person."

The Average Person

Adolphe Quetelet was born on February 22, 1796, in the ancient Belgian town of Ghent. His father, a municipal officer, died when Adolphe was seven years old. Compelled to support himself early in life, Quetelet started to teach mathematics at the young age of seventeen. When not on duty as an instructor, he composed poetry, wrote the libretto for an opera, participated in the writing of two dramas, and translated a few literary works. Still, his favorite subject remained mathematics, and he was the first to graduate with the degree of doctor of science from the University of Ghent. In 1820, Quetelet was elected as a member of the Royal Academy of Sciences in Brussels, and within a short time he became the academy's most active participant. The next few years were devoted mostly to teaching and to the publication of a few treatises on mathematics, physics, and astronomy.

Quetelet used to open his course on the history of science with the following insightful observation: "The more advanced the sciences become, the more they have tended to enter the domain of mathematics, which is a sort of center towards which they converge. We can judge of the perfection to which a science has come by the facility, more or less great, with which it may be approached by calculation."

In December of 1823, Quetelet was sent to Paris at the state's expense, mostly to study observational techniques in astronomy. As it turned out, however, this three-month visit to the then mathematical capital of the world veered Quetelet in an entirely different direction—the theory of probability. The person who was mostly responsible for igniting Quetelet's enthusiastic interest in this subject was Laplace himself. Quetelet later summarized his experience with statistics and probability:

> Chance, that mysterious, much abused word, should be considered only a veil for our ignorance; it is a phantom which exercises the most absolute empire over the common mind, accustomed to consider events only as isolated, but which is reduced to naught before the philosopher, whose eye embraces a long series of events and whose penetration is not led astray

by variations, which disappear when he gives himself sufficient perspective to seize the laws of nature.

The importance of this conclusion cannot be overemphasized. Quetelet essentially denied the role of chance and replaced it with the bold (even though not entirely proven) inference that even social phenomena have causes, and that the regularities exhibited by statistical results can be used to uncover the rules underlying social order.

In an attempt to put his statistical approach to the test, Quetelet started an ambitious project of collecting thousands of measurements related to the human body. For instance, he studied the distributions of the chest measurements of 5,738 Scottish soldiers and of the heights of 100,000 French conscripts by plotting separately the frequency with which each human trait occurred. In other words, he represented graphically how many conscripts had heights between, say, five feet and five feet two inches, and then between five feet two inches and five feet four inches, and so on. He later constructed similar curves even for what he called "moral" traits for which he had sufficient data. The latter qualities included suicides, marriages, and the propensity to crime. To his surprise, Quetelet discovered that all the human characteristics followed what is now known as the *normal* (or *Gaussian,* named somewhat unjustifiably after the "prince of mathematics" Carl Friedrich Gauss), bell-shaped frequency distribution (figure 33). Whether it was heights, weights, measurements of limb lengths, or even intellectual qualities determined by what were then pioneering psychological tests, the same type of curve appeared again and again. The curve itself was not new to Quetelet—mathematicians and physicists recognized it from the mid-eighteenth century, and Quetelet was familiar with it from his astronomical work—it was just

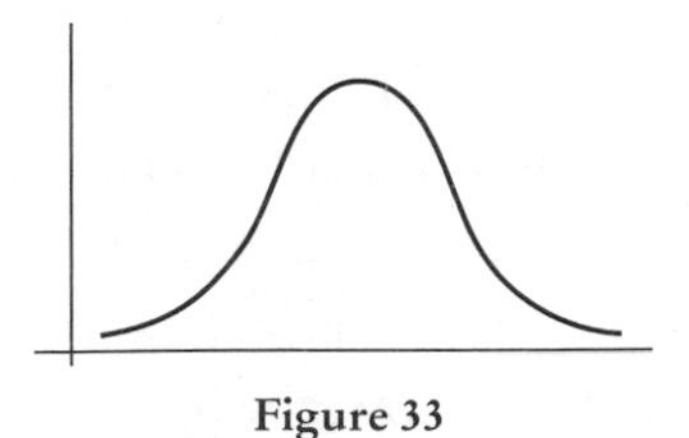

Figure 33

the association of this curve with human characteristics that came as somewhat of a shock. Previously, this curve had been known as the *error curve,* because of its appearance in any type of errors in measurements.

Imagine, for instance, that you are interested in measuring very accurately the temperature of a liquid in a vessel. You can use a high-precision thermometer and over a period of one hour take one thousand consecutive readings. You will find that due to random errors and possibly some fluctuations in the temperature, not all measurements will give precisely the same value. Rather, the measurements would tend to cluster around a central value, with some measurements giving temperatures that are higher and others that are lower. If you plot the number of times that each measurement occurred against the value of the temperature, you will obtain the same type of bell-shaped curve that Quetelet found for the human characteristics. In fact, the larger the number of measurements performed on any physical quantity, the closer will the obtained frequency distribution approximate the normal curve. The immediate implication of this fact for the question of the unreasonable effectiveness of mathematics is quite dramatic in itself—even human errors obey some strict mathematical rules.

Quetelet thought that the conclusions were even more far-reaching. He regarded the finding that human characteristics followed the error curve as an indication that the "average man" was in fact a type that nature was trying to produce. According to Quetelet, just as manufacturing errors would create a distribution of lengths around the average (correct) length of a nail, nature's errors were distributed around a preferred biological type. He declared that the people of a nation were clustered about their average "as if they were the results of measurements made on one and the same person, but with instruments clumsy enough to justify the size of the variation."

Clearly, Quetelet's speculations went a bit too far. While his discovery that biological characteristics (whether physical or mental) are distributed according to the normal frequency curve was extremely important, this could neither be taken as proof for nature's intentions nor could individual variations be treated as mere mistakes. For instance, Quetelet found the average height of the French conscripts to be five feet four inches. At the low end, however, he found a man of one foot five inches. Obviously one could not make an error of almost four feet in measuring the height of a man five feet four inches tall.

Even if we ignore Quetelet's notion of "laws" that fashion humans in a single mold, the fact that the distributions of a variety of traits ranging from weights to IQ levels all follow the normal curve is in itself pretty remarkable. And if that is not enough, even the distribution of major-league batting averages in baseball is reasonably normal, as is the annual rate of return on stock indexes (which are composed of many individual stocks). Indeed, distributions that deviate from the normal curve sometimes call for a careful examination. For instance, if the distribution of the grades in English in some school were found not to be normal, this could provoke an investigation into the grading practices of that school. This is not to say that all distributions are normal. The distribution of the lengths of words that Shakespeare used in his plays is not normal. He used many more words of three and four letters than words of eleven or twelve letters. The annual household income in the United States is also represented by a non-normal distribution. In 2006, for instance, the top 6.37% of households earned roughly one third of all income. This fact raises an interesting question in itself: If both the physical and the intellectual characteristics of humans (which presumably determine the potential for income) are normally distributed, why isn't the income? The answer to such socioeconomic questions is, however, beyond the scope of the present book. From our present limited perspective, the amazing fact is that essentially all the physically measurable particulars of humans, or of animals and plants (of any given variety) are distributed according to just one type of mathematical function.

Human characteristics served historically not only as the basis for the study of the statistical frequency distributions, but also for the establishment of the mathematical concept of *correlation*. The correlation measures the degree to which changes in the value of one variable are accompanied by changes in another. For instance, taller women may be expected to wear larger shoes. Similarly, psychologists found a correlation between the intelligence of parents and the degree to which their children succeed in school.

The concept of a correlation becomes particularly useful in those situations in which there is no precise functional dependence between the two variables. Imagine, for example, that one variable is the maximal daytime temperature in southern Arizona and the other is the number of forest fires in that region. For a given value of the temperature, one cannot predict precisely the number of forest fires that will break out, since the latter depends on other variables such as the humidity and the number of fires started by people. In other words, for any value of the temperature, there could be many corresponding numbers of forest fires and vice versa. Still, the mathematical concept known as the *correlation coefficient* allows us to measure quantitatively the strength of the relationship between two such variables.

The person who first introduced the tool of the correlation coefficient was the Victorian geographer, meteorologist, anthropologist, and statistician Sir Francis Galton (1822–1911). Galton—who was, by the way, the half-cousin of Charles Darwin—was not a professional mathematician. Being an extraordinarily practical man, he usually left the mathematical refinements of his innovative concepts to other mathematicians, in particular to the statistician Karl Pearson (1857–1936). Here is how Galton explained the concept of correlation:

> The length of the cubit [the forearm] is correlated with the stature, because a long cubit usually implies a tall man. If the correlation between them is very close, a very long cubit would usually imply a very tall stature, but if it were not very close, a very long cubit would be on the average associated with only a tall stature, and not a very tall one; while, if it were *nil,* a very long cubit would be associated with no especial stature, and therefore, on the average, with mediocrity.

Pearson eventually gave a precise mathematical definition of the correlation coefficient. The coefficient is defined in such a way that when the correlation is very high—that is, when one variable closely follows the up-and-down trends of the other—the coefficient takes the value of 1. When two quantities are *anticorrelated,* meaning that when one increases the other decreases and vice versa, the coefficient is equal to –1. Two variables that each behave as if the other didn't even exist have a correlation coefficient of 0. (For instance, the behavior of some governments unfortunately shows almost zero correlation with the wishes of the people whom they supposedly represent.)

Modern medical research and economic forecasting depend crucially on identifying and calculating correlations. The links between smoking and lung cancer, and between exposure to the Sun and skin cancer, for instance, were established initially by discovering and evaluating correlations. Stock market analysts are constantly trying to find and quantify correlations between market behavior and other variables; any such discovery can be enormously profitable.

As some of the early statisticians readily realized, both the collection of statistical data and their interpretation can be very tricky and should be handled with the utmost care. A fisherman who uses a net with holes that are ten inches on a side might be tempted to conclude that all fish are larger than ten inches, simply because the smaller ones would escape from his net. This is an example of *selection effects*—biases introduced in the results due to either the apparatus used for collecting the data or the methodology used to analyze them. Sampling presents another problem. For instance, modern opinion polls usually interview no more than a few thousand people. How can the pollsters be sure that the views expressed by members of this sample correctly represent the opinions of hundreds of millions? Another point to realize is that correlation does not necessarily imply causation. The sales of new toasters may be on the rise at the same time that audiences at concerts of classical music increase, but this does not mean that the presence of a new toaster at home enhances musical appreciation. Rather, both effects may be caused by an improvement in the economy.

In spite of these important caveats, statistics have become one of the most effective instruments in modern society, literally putting the "science" into the social sciences. But why do statistics work at all? The answer is given by the mathematics of *probability,* which reigns over many facets of modern life. Engineers trying to decide which safety mechanisms to install into the Crew Exploration Vehicle for astronauts, particle physicists analyzing results of accelerator experiments, psychologists rating children in IQ tests, drug companies evaluating the efficacy of new medications, and geneticists studying human heredity all have to use the mathematical theory of probability.

Games of Chance

The serious study of probability started from very modest beginnings—attempts by gamblers to adjust their bets to the odds of success. In particular, in the middle of the seventeenth century, a French nobleman—the Chevalier de Méré—who was also a reputed gamester, addressed a series of questions about gambling to the famous French mathematician and philosopher Blaise Pascal (1623–62). The latter conducted in 1654 an extensive correspondence about these questions with the other great French mathematician of the time, Pierre de Fermat (1601–65). The theory of probability was essentially born in this correspondence.

Let's examine one of the fascinating examples discussed by Pascal in a letter dated July 29, 1654. Imagine two noblemen engaged in a game involving the roll of a single die. Each player has put on the table thirty-two pistoles of gold. The first player chose the number 1, and the second chose the number 5. Each time the chosen number of one of the players turns up, that player gets one point. The winner is the first one to have three points. Suppose, however, that after the game has been played for some time, the number 1 has turned up twice (so that the player who had chosen that number has two points), while the number 5 has turned up only once (so the opponent has only one point). If, for whatever reason, the game has to be interrupted at that point, how should the sixty-four pistoles on the table be divided between the two players? Pascal and Fermat found the mathematically logical answer. If the player with two points were to win the next roll, the sixty-four pistoles would belong to him. If the other player were to win the next roll, each player would have had two points, and so each would have gotten thirty-two pistoles. Therefore, if the players separate without playing the next roll, the first player could correctly argue: "I am certain of thirty-two pistoles even if I lose this roll, and as for the other thirty-two pistoles perhaps I shall have them and perhaps you will have them; the chances are equal. Let us then divide these thirty-two pistoles equally and give me also the thirty-two pistoles of which I am certain." In other words, the first player should get forty-eight pistoles and the other sixteen pistoles. Unbelievable, isn't it, that a new, deep mathematical discipline could have emerged from this type of apparently trivial discussion? This is, however, precisely the reason why the effectiveness of mathematics is as "unreasonable" and mysterious as it is.

The essence of probability theory can be gleaned from the following simple facts. No one can predict with certainty which face a fair coin tossed into the air will show once it lands. Even if the coin has just come up heads ten times in a row, this does not improve our ability to predict with certainty the next toss by one iota. Yet we can predict with certainty that if you toss that coin ten million times, very close to half the tosses will show heads and very close to half will show tails. In fact, at the end of the nineteenth century, the statistician Karl Pearson had the patience to toss a coin 24,000 times. He obtained heads in 12,012 of the tosses. This is, in some sense, what probability theory is really all about. Probability theory provides us with accurate information about the collection of the results of a large number of experiments; it can never predict the result of any specific experiment. If an experiment can produce n possible outcomes, each one having the same chance of occurring, then the probability for each outcome is $1/n$. If you roll a fair die, the probability of obtaining the number 4 is 1/6, because the die has six faces, and each face is an equally likely outcome. Suppose you rolled the die seven times in a row and each time you got a 4, what would be the probability of getting a 4 in the next throw? Probability theory gives a crystal-clear answer: The probability would still be 1/6—the die has no memory and any notions of a "hot hand" or of the next roll making up for the previous imbalance are only myths. What is true is that if you were to roll the die a million times, the results will average out and 4 would appear very close to one-sixth of the time.

Let's examine a slightly more complex situation. Suppose you simultaneously toss three coins. What is the probability of getting two tails and one head? We can find the answer simply by listing all the possible outcomes. If we denote heads by "H" and tails by "T," then there are eight possible outcomes: TTT, TTH, THT, THH, HTT, HTH, HHT, HHH. Of these, you can check that three are favorable to the event "two tails and one head." Therefore, the probability for this event is 3/8. Or more generally, if out of n outcomes of equal chances, m are favorable to the event you are interested in, then the probability for that event to happen is m/n. Note that this means that the probability always takes a value between zero and one. If the event you are interested in is in fact impossible, then $m = 0$ (no outcome is favorable) and the probability would be zero. If, on the other hand, the event is absolutely certain, that means that all n events are favorable ($m = n$) and the probability is then simply $n/n = 1$. The results of the three coin tosses demonstrate yet another important result of probability theory—if you have several events that are entirely *independent* of each other, then the probability of all of them happening is the product of the individual probabilities. For instance, the probability of obtaining three heads is 1/8, which is the product of the three probabilities of obtaining heads in each of the three coins: $1/2 \times 1/2 \times 1/2 = 1/8$.

OK, you may think, but other than in casino games and other gambling activities, what additional uses can we make of these very basic probability concepts? Believe it or not, these seemingly insignificant probability laws are at the heart of the modern study of genetics—the science of the inheritance of biological characteristics.

The person who brought probability into genetics was a Moravian priest. Gregor Mendel (1822–84) was born in a village near the border between Moravia and Silesia (today Hynčice in the Czech Republic). After entering the Augustinian Abbey of St. Thomas in Brno, he studied zoology, botany, physics, and chemistry at the University of Vienna. Upon returning to Brno, he began an active experimentation with pea plants, with strong support from the abbot of the Augustinian monastery. Mendel focused his research on pea plants because they were easy to grow, and also because they have both male and female reproductive organs. Consequently, pea plants can be either self-pollinated or cross-pollinated with another plant.

By cross-pollinating plants that produce only green seeds with plants that produce only yellow seeds, Mendel obtained results that at first glance appeared to be very puzzling (figure 34). The first offspring generation had only yellow seeds. However, the following generation consistently had a 3:1 ratio of yellow to green seeds! From these surprising findings, Mendel was able to distill three conclusions that became important milestones in genetics:

1. The inheritance of a characteristic involves the transmittance of certain "factors" (what we call *genes* today) from parents to offspring.
2. Every offspring inherits one such "factor" from each parent (for any given trait).
3. A given characteristic may not manifest itself in an offspring but it can still be passed on to the following generation.

But how can one explain the quantitative results in Mendel's experiment? Mendel argued that each of the parent plants must have had two identical "factors" (what we would call alleles, varieties of a gene), either two yellow or two green (as in figure 35). When the two were mated, each offspring inherited two different alleles, one from each parent (according to rule 2 above). That is, each offspring seed contained a yellow allele and a green allele. Why then were the peas of this generation all yellow? Because, Mendel explained, yellow was the dominant color and it masked the presence of the green allele in this generation (rule 3 above). However (still according to rule 3), the dominant yellow did not prevent the recessive green from being

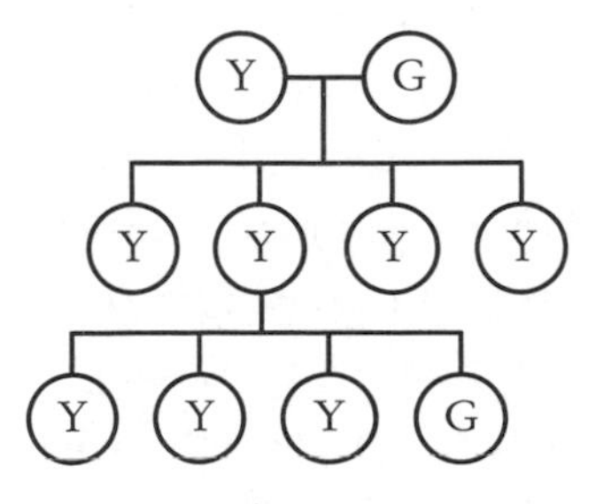

Figure 34

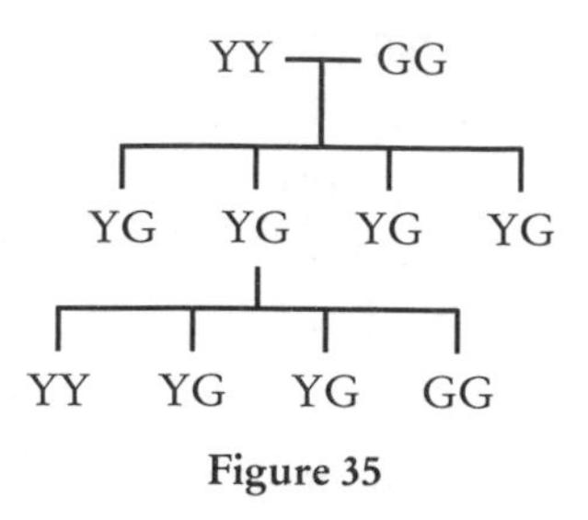

Figure 35

passed on to the next generation. In the next mating round, each plant containing one yellow allele and one green allele was pollinated with another plant containing the same combination of alleles. Since the offspring contain one allele from each parent, the seeds of the next generation may contain one of the following combinations (figure 35): green-green, green-yellow, yellow-green, or yellow-yellow. All the seeds with a yellow allele become yellow peas, because yellow is dominant. Therefore, since all the allele combinations are equally likely, the ratio of yellow to green peas should be 3:1.

You may have noticed that the entire Mendel exercise is essentially identical to the experiment of tossing two coins. Assigning heads to green and tails to yellow and asking what fraction of the peas would be yellow (given that yellow is dominant in determining the color) is precisely the same as asking what is the probability of obtaining at least one tails in tossing two coins. Clearly that is 3/4, since three of the possible outcomes (tails-tails, tails-heads, heads-tails, heads-heads) contain a tails. This means that the ratio of the number of tosses that do contain at least one tails to the number of tosses that do not should be (in the long run) 3:1, just as in Mendel's experiments.

In spite of the fact that Mendel published his paper "Experiments on Plant Hybridization" in 1865 (and he also presented the results at two scientific meetings), his work went largely unnoticed until it was rediscovered at the beginning of the twentieth century. While some questions related to the accuracy of his results have been raised, he is still regarded as the first to have laid the mathematical foundations of modern genetics. Following in the path cleared by Mendel, the influential British statistician Ronald Aylmer Fisher (1890–1962) established the field of population genetics—the mathematical branch that centers on modeling the distribution of genes within a population and on calculating how gene frequencies change over time. Today's geneticists can use statistical samplings in combination with DNA studies to forecast probable characteristics of unborn offspring. But still, how exactly are probability and statistics related?

Facts and Forecasts

Scientists who try to decipher the evolution of the universe usually try to attack the problem from both ends. There are those who start from the tiniest fluctuations in the cosmic fabric in the primordial universe, and there are those who study every detail in the current state of the universe. The former use large computer simulations in an attempt to evolve the universe forward. The latter engage in the detective-style work of trying to deduce the universe's past from a multitude of facts about its present state. Probability theory and statistics are related in a similar fashion. In probability theory the variables and the initial state are known, and the goal is to predict the most likely end result. In statistics the outcome is known, but the past causes are uncertain.

Let's examine a simple example of how the two fields supplement each other and meet, so to speak, in the middle. We can start from the fact that statistical studies show that the measurements of a large variety of physical quantities and even of many human characteristics are distributed according to the *normal frequency curve*. More precisely, the normal curve is not a single curve, but rather a family of curves, all describable by the same general function, and all being fully characterized by just two mathematical quantities. The first of these quantities—the *mean*—is the central value about which the distribution is symmetric. The actual value of the mean depends, of course, on the type of variable being measured (e.g., weight, height, or IQ). Even for the same variable, the mean may be different for different populations. For instance, the mean of the heights of men in Sweden is probably different from the mean of the heights of men in Peru. The second quantity that defines the normal curve is known as the *standard deviation*. This is a measure of how closely the data are clustered around the mean value. In figure 36, the normal curve (a) has the largest standard

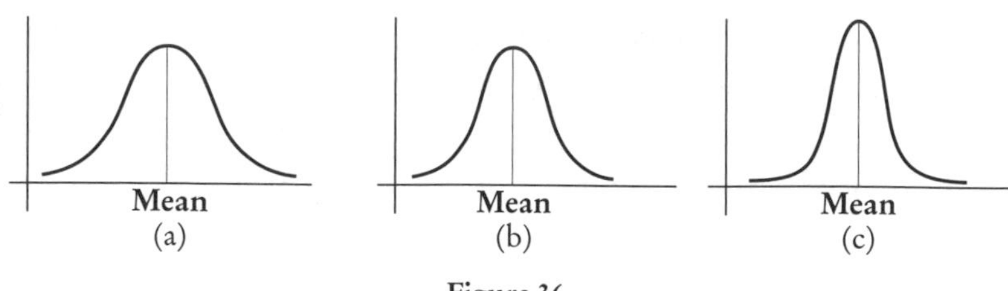

Figure 36

deviation, because the values are more widely dispersed. Here, however, comes an interesting fact. By using integral calculus to calculate areas under the curve, one can prove mathematically that irrespective of the values of the mean or the standard deviation, 68.2 percent of the data lie within the values encompassed by one standard deviation on either side of the mean (as in figure 37). In other words, if the mean IQ of a certain (large) population is 100, and the standard deviation is 15, then 68.2 percent of the people in that population have IQ values between 85 and 115. Furthermore, for all the normal frequency curves, 95.4 percent of all the cases lie within two standard deviations of the mean, and 99.7 percent of the data lie within three standard deviations on either side of the mean (figure 37). This implies that in the above example, 95.4 percent of the population have IQ values between 70 and 130, and 99.7 percent have values between 55 and 145.

Suppose now that we want to predict what the probability would be for a person chosen at random from that population to have an IQ value between 85 and 100. Figure 37 tells us that the probability would be 0.341 (or 34.1 percent), since according to the laws of probability, the probability is simply the number of favorable outcomes divided by the total number of possibilities. Or we could be interested in finding out what the probability is for someone (chosen at random) to have an IQ value higher than 130 in that population. A glance at figure 37 reveals that the probability is only about 0.022, or 2.2 percent. Much in the same way, using the properties of the normal distribution and the tool of integral calculus (to calculate areas), one can calculate the probability of the IQ value being in any given range. In other words, probability theory and its complementary helpmate, statistics, combine to give us the answer.

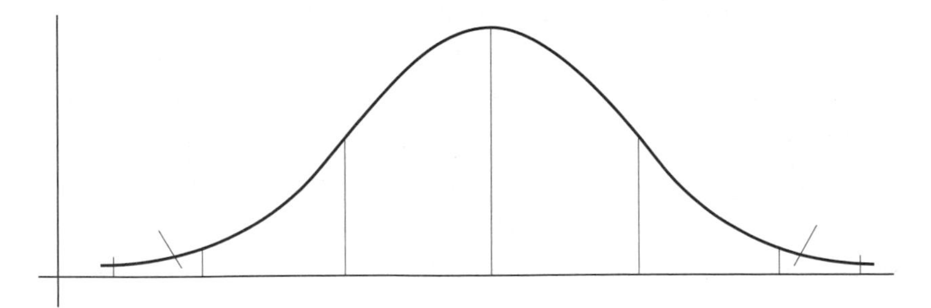

Figure 37

As I have noted several times already, probability and statistics become meaningful when one deals with a large number of events—never individual events. This cardinal realization, known as the *law of large numbers,* is due to Jakob Bernoulli, who formulated it as a theorem in his book *Ars Conjectandi* (*The Art of Conjecturing;* figure 38 shows the frontispiece). In simple terms, the theorem states

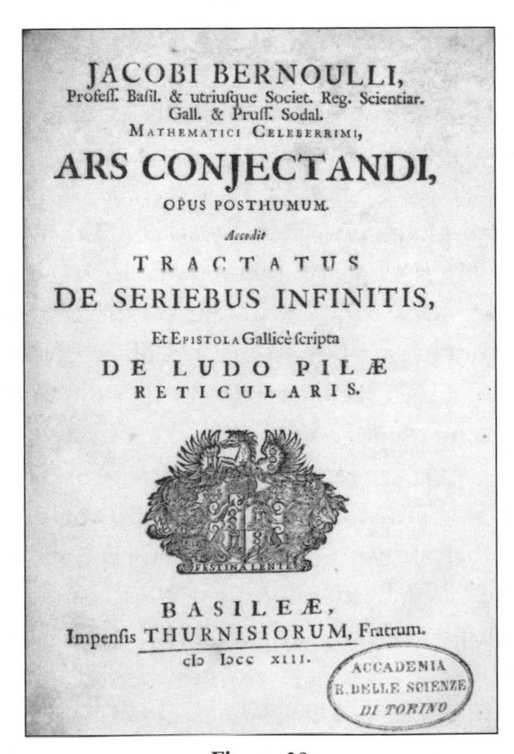

JACOBI BERNOULLI,
Profeſſ. Baſil. & utriuſque Societ. Reg. Scientiar.
Gall. & Pruſſ. Sodal.
MATHEMATICI CELEBERRIMI,

ARS CONJECTANDI,
OPUS POSTHUMUM.
Accedit
TRACTATUS
DE SERIEBUS INFINITIS,
Et EPISTOLA Gallicè ſcripta
DE LUDO PILÆ
RETICULARIS.

BASILEÆ,
Impenſis THURNISIORUM, Fratrum.
cIↄ Iↄcc xiii.

Figure 38

that if the probability of an event's occurrence is p, then p is the most probable proportion of the event's occurrences to the total number of trials. In addition, as the number of trials approaches infinity, the proportion of successes becomes p with certainty. Here is how Bernoulli introduced the law of large numbers in *Ars Conjectandi:* "What is still to be investigated is whether by increasing the number of observations we thereby also keep increasing the probability that the recorded proportion of favorable to unfavorable instances will approach the true ratio, so that this probability will finally exceed any desired degree of certainty." He then proceeded to explain the concept with a specific example:

> We have a jar containing 3000 small white pebbles and 2000 black ones, and we wish to determine empirically the ratio of white pebbles to the black—something we do not know—by drawing one pebble after another out of the jar, and recording how often a white pebble is drawn and how often a black. (I remind you that an important requirement of this process is that you put back each pebble, after noting the color, before drawing the next one, so that the number of pebbles in the urn remains constant.) Now we ask, is it possible by indefinitely extending the trials to make it 10, 100, 1000, etc., times more probable (and ultimately "morally certain") that the ratio of the number of drawings of a white pebble to the number of drawings of a black pebble will take on the same value (3:2) as the actual ratio of white to black pebbles in the urn, than that the ratio of the drawings will take on a different value? If the answer is no, then I admit that we are likely to fail in the attempt to ascertain the number of instances of each case (i.e., the number of white and of black pebbles) by observation. But if it is true that we can finally attain moral certainty by this method [and Jakob Bernoulli proves this to be the case in the following chapter of *Ars Conjectandi*] . . . then we can determine the number of instances *a posteriori* with almost as great accuracy as if they were known to us *a priori.*

Bernoulli devoted twenty years to the perfection of this theorem, which has since become one of the central pillars of statistics. He concluded with his belief in the ultimate existence of governing laws, even in those instances that appear to be a matter of chance:

> If all events from now through eternity were continually observed (whereby probability would ultimately become certainty), it would be found that everything in the world occurs for definite reasons and in definite conformity with law, and that hence we are constrained, even for things that may seem quite accidental, to assume a certain necessity and, as it were, fatefulness. For all I know that is what Plato had in mind when, in the doctrine of the universal cycle, he maintained that after the passage of countless centuries everything would return to its original state.

The upshot of this tale of the science of uncertainty is very simple: Mathematics is applicable in some ways even in the less "scientific" areas of our lives—including those that appear to be governed by pure chance. So in attempting to explain the "unreasonable effectiveness" of mathematics we cannot limit our discussion only to the laws of physics. Rather, we will eventually have to somehow figure out what it is that makes mathematics so omnipresent.

The incredible powers of mathematics were not lost on the famous playwright and essayist George Bernard Shaw (1856–1950). Definitely not known for his mathematical talents, Shaw once wrote an insightful article about statistics and probability entitled "The Vice of Gambling and the Virtue of Insurance." In this article, Shaw admits that to him insurance is "founded on facts that are inexplicable and risks that are calculable only by professional mathematicians." Yet he offers the following perceptive observation:

> Imagine then a business talk between a merchant greedy for foreign trade but desperately afraid of being shipwrecked or eaten by savages, and a skipper greedy for cargo and passen-

gers. The captain answers the merchant that his goods will be perfectly safe, and himself equally so if he accompanies them. But the merchant, with his head full of the adventures of Jonah, St. Paul, Odysseus, and Robinson Crusoe, dares not venture. Their conversation will be like this:

Captain: Come! I will bet you umpteen pounds that if you sail with me you will be alive and well this day a year.

Merchant: But if I take the bet I shall be betting you that sum that I shall die within the year.

Captain: Why not if you lose the bet, as you certainly will?

Merchant: But if I am drowned you will be drowned too; and then what becomes of our bet?

Captain: True. But I will find you a landsman who will make the bet with your wife and family.

Merchant: That alters the case of course; but what about my cargo?

Captain: Pooh! The bet can be on the cargo as well. Or two bets: one on your life, the other on the cargo. Both will be safe, I assure you. Nothing will happen; and you will see all the wonders that are to be seen abroad.

Merchant: But if I and my goods get through safely I shall have to pay you the value of my life and of the goods into the bargain. If I am not drowned I shall be ruined.

Captain: That also is very true. But there is not so much for me in it as you think. If you are drowned I shall be drowned first; for I must be the last man to leave the sinking ship. Still, let me persuade you to venture. I will make the bet ten to one. Will that tempt you?

Merchant: Oh, in that case—

The captain has discovered insurance just as the goldsmiths discovered banking.

For someone such as Shaw, who complained that during his education "not a word was said to us about the meaning or utility of mathematics," this humorous account of the "history" of the mathematics of insurance is quite remarkable.

With the exception of Shaw's text, we have so far followed the development of some branches of mathematics more or less through the eyes of practicing mathematicians. To these individuals, and indeed to many rationalist philosophers such as Spinoza, Platonism was obvious. There was no question that mathematical truths existed in their own world and that the human mind could access these verities without any observation, solely through the faculty of reason. The first signs of a potential gap between the perception of Euclidean geometry as a collection of universal truths and other branches of mathematics were uncovered by the Irish philosopher George Berkeley, Bishop of Cloyne (1685–1753). In a pamphlet entitled *The Analyst; Or a Discourse Addressed to An Infidel Mathematician* (the latter presumed to be Edmond Halley), Berkeley criticized the very foundations of the fields of calculus and analysis, as introduced by Newton (in *Principia*) and Leibniz. In particular, Berkeley demonstrated that Newton's concept of "fluxions," or instantaneous rates of change, was far from being rigorously defined, which in Berkeley's mind was sufficient to cast doubt on the entire discipline:

> The method of fluxions is the general key, by help whereof the modern Mathematicians unlock the secrets of Geometry, and consequently of Nature . . . But whether this Method be clear or obscure, consistent or repugnant, demonstrative or precarious, as I shall inquire with the utmost impartiality, so I submit my inquiry to your own Judgement, and that of every candid Reader.

Berkeley certainly had a point, and the fact is that a fully consistent theory of analysis was only formulated in the 1960s. But mathematics was about to experience a more dramatic crisis in the nineteenth century.

CHAPTER 6

GEOMETERS: FUTURE SHOCK

In his famous book *Future Shock,* author Alvin Toffler defined the term in the title as "the shattering stress and disorientation that we induce in individuals by subjecting them to too much change in too short a time." In the nineteenth century, mathematicians, scientists, and philosophers experienced precisely such a shock. In fact, the millennia-old belief that mathematics offers eternal and immutable truths was crushed. This unexpected intellectual upheaval was caused by the emergence of new types of geometries, now known as *non-Euclidean geometries.* Even though most nonspecialists may have never even heard of non-Euclidean geometries, the magnitude of the revolution in thought introduced by these new branches of mathematics has been likened by some to that inaugurated by the Darwinian theory of evolution.

To fully appreciate the nature of this sweeping change in worldview, we have first to briefly probe the historical-mathematical backdrop.

Euclidean "Truth"

Until the beginning of the nineteenth century, if there was one branch of knowledge that had been regarded as the apotheosis of truth and certainty, it was Euclidean geometry, the traditional geometry we learn in school. Not surprisingly, therefore, the great Dutch Jewish philosopher Baruch Spinoza (1632–77) entitled his bold attempt to unify science, religion, ethics, and reason *Ethics, Demonstrated in*

Geometrical Order. Moreover, in spite of the clear distinction between the ideal, Platonic world of mathematical forms and physical reality, most scientists regarded the objects of Euclidean geometry simply as the distilled abstractions of their real, physical counterparts. Even staunch empiricists such as David Hume (1711–76), who insisted that the very foundations of science were far less certain than anyone had ever suspected, concluded that Euclidean geometry was as solid as the Rock of Gibraltar. In *An Enquiry Concerning Human Understanding,* Hume identified "truths" of two types:

> All the objects of human reason or enquiry may naturally be divided into two kinds, to wit, Relations of Ideas, and Matters of Fact. Of the first kind are . . . every affirmation which is either intuitively or demonstratively certain . . . Propositions of this kind are discoverable by the mere operation of thought, without dependence on what is anywhere existent in the universe. Though there never were a circle or triangle in nature, the truths demonstrated by Euclid would forever retain their certainty and evidence. Matters of fact . . . are not ascertained in the same manner; nor is our evidence of their truth, however great, of a like nature with the foregoing. The contrary of every matter of fact is still possible; because it can never imply a contradiction . . . That the sun will not rise tomorrow is no less intelligible a proposition, and implies no more contradiction than the affirmation, that it will rise. We should in vain, therefore, attempt to demonstrate its falsehood.

In other words, while Hume, like all empiricists, maintained that all knowledge stems from observation, geometry and its "truths" continued to enjoy a privileged status.

The preeminent German philosopher Immanuel Kant (1724–1804) did not always agree with Hume, but he also exalted Euclidean geometry to a status of absolute certainty and unquestionable validity. In his memorable *Critique of Pure Reason,* Kant attempted to reverse in some sense the relationship between the mind and the physical world. Instead of impressions of physical reality being imprinted on an otherwise entirely passive mind, Kant gave the mind the active function of "constructing" or "processing" the perceived universe. Turning his attention inward, Kant asked not *what* we can know, but *how* we can know what we can know. He explained that while our eyes detect particles of light, these do not form an image in our awareness until the information is processed and organized by our brains. A key role in this construction process was assigned to the human intuitive or synthetic *a priori* grasp of space, which in turn was taken to be based on Euclidean geometry. Kant believed that Euclidean geometry provided the only true path for processing and conceptualizing space, and that this intuitive, universal acquaintance with space was at the heart of our experience of the natural world. In Kant's words:

> Space is not an empirical concept which has been derived from external experience . . . Space is a necessary representation *a priori,* forming the very foundation of all external intuitions . . . On this necessity of an *a priori* representation of space rests the apodictic certainty of all geometrical principles, and the possibility of their construction *a priori.* For if the intuition of space were a concept gained *a posteriori,* borrowed from general external experience, the first principles of mathematical definition would be nothing but perceptions. They would be exposed to all the accidents of perception, and there being but one straight line between two points would not be a necessity, but only something taught in each case by experience.

To put it simply, according to Kant, if we perceive an object, then necessarily this object is spatial and Euclidean.

Hume's and Kant's ideas bring to the forefront the two rather different, but equally important aspects that had been historically associated with Euclidean geometry. The first was the statement that Euclidean geometry represents the only accurate description of physical space. The second was the identification of Euclidean geometry with a firm, decisive, and infallible deductive structure. Taken together, these two presumed properties provided mathematicians, scientists, and philosophers with what they regarded as the strongest evidence that informative, inescapable truths about the universe do exist. Until the nineteenth century these statements were taken for granted. But were they actually true?

The foundations of Euclidean geometry were laid around 300 BC by the Greek mathematician Euclid of Alexandria. In a monumental thirteen-volume opus entitled *The Elements,* Euclid attempted to erect geometry on a well-defined logical base. He started with ten axioms assumed to be indisputably true and sought to prove a large number of propositions on the basis of those postulates by nothing other than logical deductions.

The first four Euclidean axioms were extremely simple and exquisitely concise. For instance, the first axiom read: "Between any two points a straight line may be drawn." The fourth one stated: "All right angles are equal." By contrast, the fifth axiom, known as the "parallel postulate," was more complicated in its formulation and considerably less self-evident: "If two lines lying in a plane intersect a third line in such a way that the sum of the internal angles on one side is less than the two right angles, then the two lines inevitably will intersect each other if extended sufficiently on that side." Figure 39 demonstrates graphically the contents of this axiom. While no one doubted the truth of this statement, it lacked the compelling simplicity of the other axioms. All indications are that even Euclid himself was not entirely happy with his fifth postulate— the proofs of the

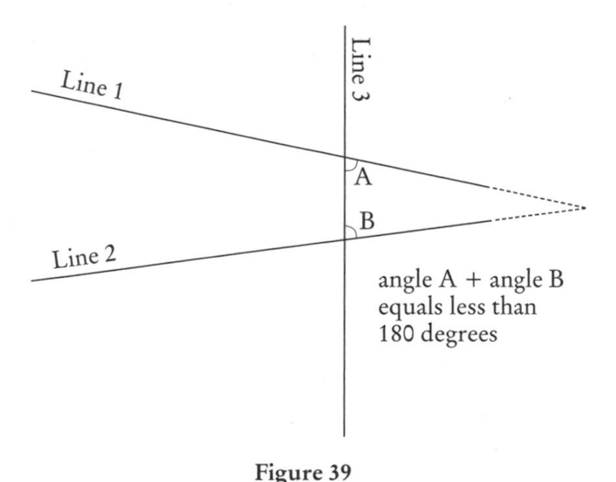

Figure 39

first twenty-eight propositions in *The Elements* do not make use of it. The equivalent version of the "fifth" most cited today appeared first in commentaries by the Greek mathematician Proclus in the fifth century, but it is generally known as the "Playfair axiom," after the Scottish mathematician John Playfair (1748–1819). It states: "Given a line and a point not on the line, it is possible to draw exactly one line parallel to the given line through that point" (see figure 40). The two versions of the axiom are equivalent in the sense that Playfair's axiom (together with the other axioms) necessarily implies Euclid's original fifth axiom and vice versa.

Over the centuries, the increasing discontent with the fifth axiom resulted in a number of unsuccessful attempts to actually prove it from the other nine axioms or to replace it by a more obvious postulate. When those efforts failed, other geometers began trying to answer an intriguing "what if" question—what if the fifth axiom did, in fact, not prove true? Some of those endeavors started to raise nagging doubts on whether Euclid's axioms were truly self-evident, rather than being based on experience. The final, surprising verdict eventually came in the nineteenth century: One could create new kinds of geometry by *choosing* an axiom different from Euclid's fifth. Furthermore, these "non-Euclidean" geometries could in principle describe physical space just as accurately as Euclidean geometry did!

Let me pause here for a moment to allow for the meaning of the word "choosing" to sink in. For millennia, Euclidean geometry had been regarded as unique and *inevitable*—the sole true description of space. The fact that one could choose the axioms and obtain an equally valid description turned the entire concept on its ear. The certain, carefully constructed deductive scheme suddenly became more similar to a game, in which the axioms simply played the role of the rules. You could change the axioms and play a different game. The

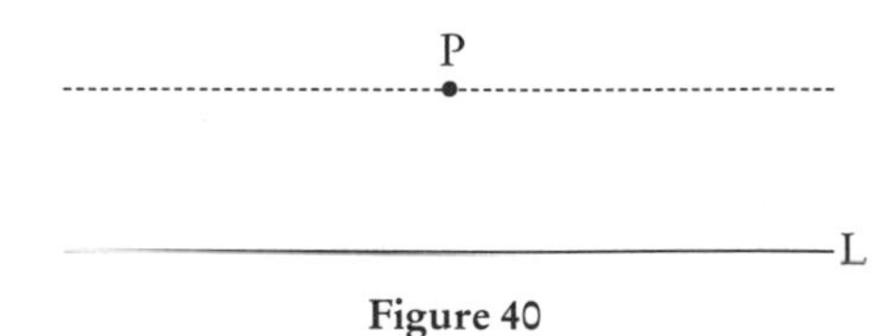

Figure 40

impact of this realization on the understanding of the nature of mathematics cannot be overemphasized.

Quite a few creative mathematicians prepared the ground for the final assault on Euclidean geometry. Particularly notable among them were the Jesuit priest Girolamo Saccheri (1667–1733), who investigated the consequences of replacing the fifth postulate by a different statement, and the German mathematicians Georg Klügel (1739–1812) and Johann Heinrich Lambert (1728–1777), who were the first to realize that alternative geometries to the Euclidean could exist. Still, somebody had to put the last nail in the coffin of the idea of Euclidean geometry being the one and only representation of space. That honor was shared by three mathematicians, one from Russia, one from Hungary, and one from Germany.

Strange New Worlds

The first to publish an entire treatise on a new type of geometry—one that could be constructed on a surface shaped like a curved saddle (figure 41a)—was the Russian Nikolai Ivanovich Lobachevsky (1792–1856; figure 42). In this kind of geometry (now known as *hyperbolic geometry*), Euclid's fifth postulate is replaced by the statement that given a line in a plane and a point not on this line, there are at least two lines through the point parallel to the given line. Another important difference between Lobachevskian geometry and Euclidean geometry is that while in the latter the angles in a triangle always add up to 180 degrees (figure 41b), in the former the sum is always

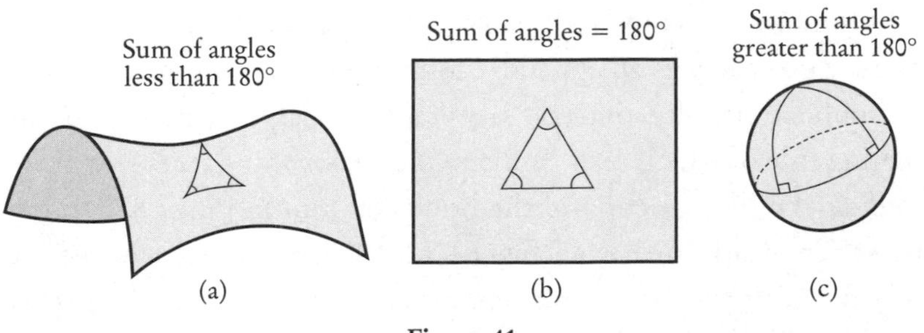

Figure 41

Figure 42

less than 180 degrees. Because Lobachevsky's work appeared in the rather obscure *Kazan Bulletin,* it went almost entirely unnoticed until French and German translations started to appear in the late 1830s. Unaware of Lobachevsky's work, a young Hungarian mathematician, János Bolyai (1802–60), formulated a similar geometry during the 1820s. Bursting with youthful enthusiasm, he wrote in 1823 to his father (the mathematician Farkas Bolyai; figure 43): "I have found things so magnificent that I was astounded . . . I have created a different new world out of nothing." By 1825, János was already able to present to the elder Bolyai the first draft of his new geometry. The manuscript was entitled *The Science Absolute of Space.* In spite of the young man's exuberance, the father was not entirely convinced of the soundness of János's ideas. Nevertheless, he decided to publish the new geometry as an appendix to his own two-volume treatise on the foundations of geometry, algebra, and analysis (the supposedly inviting title of which read *Essay on the Elements of Mathematics for Studious Youths*). A copy of the book was sent in June 1831 to Farkas's friend Carl Friedrich Gauss (1777–1855; figure 44), who was not only the most prominent mathematician of the time, but who is also considered by many, along with Archimedes and Newton, to be one

Figure 43

of the three greatest of all time. That book was somehow lost in the chaos created by a cholera epidemic, and Farkas had to send a second copy. Gauss sent out a reply on March 6, 1832, and his comments were not exactly what the young János expected:

> If I commenced by saying that I am unable to praise this work, you would certainly be surprised for a moment. But I cannot say otherwise. To praise it, would be to praise myself. Indeed the whole contents of the work, the path taken by your son, the results to which he is led, coincide almost entirely with my meditations, which have occupied my mind partly for the last thirty or thirty-five years. So I remained quite stupefied. So far as my own work is concerned, of which up till now I have put little on paper, my intention was not to let it be published during my lifetime.

Let me parenthetically note that apparently Gauss feared that the radically new geometry would be regarded by the Kantian philosophers, to whom he referred as "the Boetians" (synonymous with "stupid" for the ancient Greeks), as philosophical heresy. Gauss then continued:

Figure 44

> On the other hand it was my idea to write down all this later so that at least it should not perish with me. It is therefore a pleasant surprise with me that I am spared this trouble, and I am very glad that it is the son of my old friend, who takes the precedence of me in such a remarkable manner.

While Farkas was quite pleased with Gauss's praise, which he took to be "very fine," János was absolutely devastated. For almost a decade he refused to believe that Gauss's claim to priority was not false, and his relationship with his father (whom he suspected of prematurely communicating the results to Gauss) was seriously strained. When he finally realized that Gauss had actually started working on the problem as early as 1799, János became deeply embittered, and his subsequent mathematical work (he left some twenty thousand pages of manuscript when he died) became rather lackluster by comparison.

There is very little doubt, however, that Gauss had indeed given considerable thought to non-Euclidean geometry. In a diary entry from September 1799 he wrote: "*In principiis geometriae egregios progressus fecimus*" ("About the principles of geometry we obtained wonderful achievements"). Then, in 1813, he noted: "In the theory of parallel lines we are now no further than Euclid was. This is the *partie honteuse* [shameful part] of mathematics, which sooner or later must get a very different form." A few years later, in a letter written on April 28, 1817, he stated: "I am coming more and more to the conviction that the necessity of our [Euclidean] geometry cannot be proved." Finally, and contrary to Kant's views, Gauss concluded that Euclidean geometry could not be viewed as a universal truth, and that rather "one would have to rank [Euclidean] geometry not with arithmetic, which stands *a priori,* but approximately with mechanics." Similar conclusions were reached independently by Ferdinand Schweikart (1780–1859), a professor of jurisprudence, and the latter informed Gauss of his work sometime in 1818 or 1819. Since neither Gauss nor Schweikart actually published his results, however, the priority of first publication is traditionally credited to Lobachevsky and Bolyai, even though the two can hardly be regarded as the sole "creators" of non-Euclidean geometry.

Hyperbolic geometry broke on the world of mathematics like a thunderbolt, dealing a tremendous blow to the perception of Euclidean geometry as the only, infallible description of space. Prior to the Gauss-Lobachevsky-Bolyai work, Euclidean geometry *was,* in effect, the natural world. The fact that one could select a different set of axioms and construct a different type of geometry raised for the first time the suspicion that mathematics is, after all, a human invention, rather than a discovery of truths that exist independently of the human mind. At the same time, the collapse of the immediate connection between Euclidean geometry and true physical space exposed what appeared to be fatal deficiencies in the idea of mathematics as the language of the universe.

Euclidean geometry's privileged status went from bad to worse when one of Gauss's students, Bernhard Riemann, showed that hyperbolic geometry was not the only non-Euclidean geometry possible. In a brilliant lecture delivered in Göttingen on June 10, 1854 (figure 45 shows the first page of the published lecture), Riemann presented his views "On the Hypotheses That Lie at the Foundations of Geometry." He started by saying that "geometry presupposes the concept of space, as well as assuming the basic principles

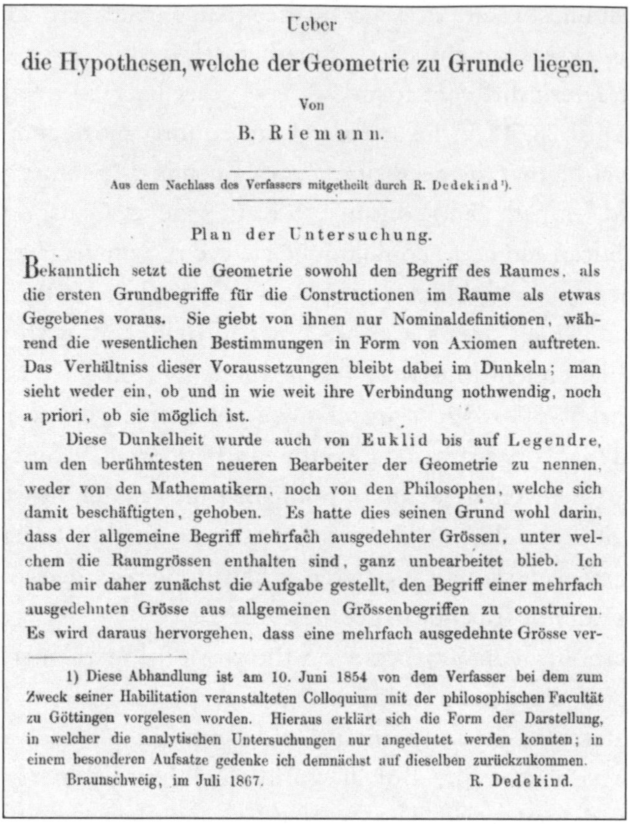

Ueber

die Hypothesen, welche der Geometrie zu Grunde liegen.

Von

B. Riemann.

Aus dem Nachlass des Verfassers mitgetheilt durch R. Dedekind[1]).

Plan der Untersuchung.

Bekanntlich setzt die Geometrie sowohl den Begriff des Raumes, als die ersten Grundbegriffe für die Constructionen im Raume als etwas Gegebenes voraus. Sie giebt von ihnen nur Nominaldefinitionen, während die wesentlichen Bestimmungen in Form von Axiomen auftreten. Das Verhältniss dieser Voraussetzungen bleibt dabei im Dunkeln; man sieht weder ein, ob und in wie weit ihre Verbindung nothwendig, noch a priori, ob sie möglich ist.

Diese Dunkelheit wurde auch von Euklid bis auf Legendre, um den berühmtesten neueren Bearbeiter der Geometrie zu nennen, weder von den Mathematikern, noch von den Philosophen, welche sich damit beschäftigten, gehoben. Es hatte dies seinen Grund wohl darin, dass der allgemeine Begriff mehrfach ausgedehnter Grössen, unter welchem die Raumgrössen enthalten sind, ganz unbearbeitet blieb. Ich habe mir daher zunächst die Aufgabe gestellt, den Begriff einer mehrfach ausgedehnten Grösse aus allgemeinen Grössenbegriffen zu construiren. Es wird daraus hervorgehen, dass eine mehrfach ausgedehnte Grösse ver-

1) Diese Abhandlung ist am 10. Juni 1854 von dem Verfasser bei dem zum Zweck seiner Habilitation veranstalteten Colloquium mit der philosophischen Facultät zu Göttingen vorgelesen worden. Hieraus erklärt sich die Form der Darstellung, in welcher die analytischen Untersuchungen nur angedeutet werden konnten; in einem besonderen Aufsatze gedenke ich demnächst auf dieselben zurückzukommen.
Braunschweig, im Juli 1867. R. Dedekind.

Figure 45

for constructions in space. It gives only nominal definitions of these things, while their essential specifications appear in the form of axioms." However, he noted, "The relationship between these presuppositions is left in the dark; we do not see whether, or to what extent, any connection between them is necessary, or *a priori* whether any connection between them is even possible." Among the possible geometrical theories Riemann discussed *elliptic geometry*, of the type that one would encounter on the surface of a sphere (figure 41c). Note that in such a geometry the shortest distance between two points is not a straight line, it is rather a segment of a great circle, whose center coincides with the center of the sphere. Airlines take advantage of this fact—flights from the United States to Europe do not follow what would appear as a straight line on the map, but rather a great circle that initially bears northward. You can easily check that any two great circles meet at two diametrically opposite points. For instance, two meridians on Earth, which appear to be parallel at the Equator, meet at the two poles. Consequently, unlike in Euclidean geometry, where there is exactly one parallel line through an external point, and hyperbolic geometry, in which there are at least two parallels, there are *no* parallel lines at all in the elliptic geometry on a sphere. Riemann took the non-Euclidean concepts one step further and introduced geometries in curved spaces in three, four, and even more dimensions. One of the key concepts expanded upon by Riemann was that of the *curvature*—the rate at which a curve or a surface curves. For instance, the surface of an eggshell curves more gently around its girth than along a curve passing through one of its pointy edges. Riemann proceeded to give a precise mathematical definition of curvature in spaces of any number of dimensions. In doing so he solidified the marriage between algebra and geometry that had been initiated by Descartes. In Riemann's work equations in any number of variables found their geometrical counterparts, and new concepts from the advanced geometries became partners of equations.

Euclidean geometry's eminence was not the only victim of the new horizons that the nineteenth century opened for geometry. Kant's ideas of space did not survive much longer. Recall that Kant asserted that information from our senses is organized exclusively along Euclidean templates before it is recorded in our consciousness. Geometers of the nineteenth century quickly developed intuition in the non-Euclidean geometries and learned to experience the world along those lines. The Euclidean perception of space turned out to be learned after all, rather than intuitive. All of these dramatic developments led the great French mathematician Henri Poincaré (1854–1912) to conclude that the axioms of geometry are "neither synthetic *a priori* intuitions nor experimental facts. They are *conventions* [emphasis added]. Our choice among all possible conventions is guided by experimental facts, but it remains free." In other words, Poincaré regarded the axioms only as "definitions in disguise."

Poincaré's views were inspired not just by the non-Euclidean geometries described so far, but also by the proliferation of other new geometries, which before the end of the nineteenth century seemed to be almost getting out of hand. In *projective geometry* (such as that obtained when an image on celluloid film is projected onto a screen), for instance, one could literally interchange the roles of points and lines, so that theorems about points and lines (in this order) became theorems about lines and points. In *differential geometry,* mathematicians used calculus to study the local geometrical properties of various mathematical spaces, such as the surface of a sphere or a torus. These and other geometries appeared, at first blush at least, to be ingenious inventions of imaginative mathematical minds, rather than accurate descriptions of physical space. How then could one still defend the concept of God as a mathematician? After all, if "God ever geometrizes" (a phrase attributed to Plato by the historian Plutarch), which of these many geometries does the divine practice?

The rapidly deepening recognition of the shortcomings of the classical Euclidean geometry forced mathematicians to take a serious look at the foundations of mathematics in general, and at the relationship between mathematics and logic in particular. We shall return to this important topic in chapter 7. Here let me only note that the very notion of the self-evidency of axioms had been shattered. Consequently, while the nineteenth century witnessed other significant developments in algebra and in analysis, the revolution in geometry probably had the most influential effects on the views of the nature of mathematics.

On Space, Numbers, and Humans

Before mathematicians could turn to the overarching topic of the foundations of mathematics, however, a few "smaller" issues required immediate attention. First, the fact that non-Euclidean geometries had been formulated and published did not necessarily mean that these were legitimate offspring of mathematics. There was the ever-present fear of inconsistency—the possibility that carrying these geometries to their ultimate logical consequences would produce unresolvable contradictions. By the 1870s, the Italian Eugenio Beltrami (1835–1900) and the German Felix Klein (1849–1925) had demonstrated that as long as Euclidean geometry was consistent, so were non-Euclidean geometries. This still left open the bigger question of the solidity of the foundations of Euclidean geometry. Then there was the important matter of relevance. Most mathematicians regarded the new geometries as amusing curiosities at best. Whereas Euclidean geometry derived much of its historical power from being seen as the description of real space, the non-Euclidean geometries had been perceived initially as not having any connection whatsoever to physical reality. Consequently, the non-Euclidean geometries were treated by many mathematicians as Euclidean geometry's poor cousins. Henri Poincaré was a bit more accommodating than most, but even he insisted that if humans were to be transported to a world in which the accepted geometry was non-Euclidean, then it was still "certain that we should not find it more convenient to make a change" from Euclidean to non-Euclidean geometry. Two questions therefore loomed large: (1) Could geometry (in particular) and other branches of mathematics (in general) be established on solid axiomatic logical foundations? and (2) What was the relationship, if any, between mathematics and the physical world?

Some mathematicians adopted a pragmatic approach with respect to the validation of the foundations of geometry. Disappointed by the realization that what they regarded as absolute truths turned out to be more experience-based than rigorous, they turned to arithmetic—the mathematics of numbers. Descartes' analytic geometry, in which points in the plane were identified with ordered pairs of numbers, circles with pairs satisfying a certain equation (see chapter 4), and so on, provided just the necessary tools for the re-erection of the foundations of geometry on the basis of numbers. The German mathematician Jacob Jacobi (1804–51) presumably expressed those shifting tides when he replaced Plato's "God ever geometrizes" by his own motto: "God ever arithmetizes." In some sense, however, these efforts only transported the problem to a different branch of mathematics. While the great German mathematician David Hilbert (1862–1943) did succeed in demonstrating that Euclidean geometry was consistent as long as arithmetic was consistent, the consistency of the latter was far from unambiguously established at that point.

On the relationship between mathematics and the physical world, a new sentiment was in the air. For many centuries, the interpretation of mathematics as a reading of the cosmos had been dramatically and continuously enhanced. The mathematization of the sciences by Galileo, Descartes, Newton, the Bernoullis, Pascal, Lagrange, Quetelet, and others was taken as strong evidence for an underlying mathematical design in nature. One could clearly argue that if mathematics wasn't the language of the cosmos, why did it work as well as it did in explaining things ranging from the basic laws of nature to human characteristics?

To be sure, mathematicians did realize that mathematics dealt only with rather abstract Platonic forms, but those were regarded as reasonable idealizations of the actual physical elements. In fact, the feeling that the book of nature was written in the language of mathematics was so deeply rooted that many mathematicians absolutely refused even to consider mathematical concepts and structures that were not directly related to the physical world. This was the case, for instance, with the colorful Gerolamo Cardano (1501–76). Cardano was an accomplished mathematician, renowned physician, and compulsive gambler. In 1545 he published one of the most influential books in the history of algebra—the *Ars Magna* (*The Great Art*). In this comprehensive treatise Cardano explored in great detail solutions to algebraic equations, from the simple quadratic equation (in which the unknown appears to the second power: x^2) to pioneering solutions to the cubic (involving x^3), and quartic (involving x^4) equations. In classical mathematics, however, quantities were often interpreted as geometrical elements. For instance, the value of the unknown x was identified with a line segment of that length, the second power x^2 was an area, and the third power x^3 was a solid having the corresponding volume. Consequently, in the first chapter of the *Ars Magna,* Cardano explains:

> We conclude our detailed consideration with the cubic, others being merely mentioned, even if generally, in passing. For as *positio* [the first power] refers to a line, *quadratum* [the square] to a surface, and *cubum* [the cube] to a solid body, it would be very foolish for us to go beyond this point. Nature

> does not permit it. Thus, it will be seen, all those matters up to and including the cubic are fully demonstrated, but the others which we will add, either by necessity or out of curiosity, we do not go beyond barely setting out.

In other words, Cardano argues that since the physical world as perceived by our senses contains only three dimensions, it would be silly for mathematicians to concern themselves with a higher number of dimensions, or with equations of a higher degree.

A similar opinion was expressed by the English mathematician John Wallis (1616–1703), from whose work *Arithmetica Infinitorum* Newton learned methods of analysis. In another important book, *Treatise of Algebra,* Wallis first proclaimed: "Nature, in propriety of Speech, doth not admit more than *three* (local) dimensions." He then elaborated:

> A Line drawn into a Line, shall make a Plane or Surface; this drawn into a Line, shall make a Solid. But if this Solid be drawn into a Line, or this Plane into a Plane, what shall it make? A Plano-Plane? This is a Monster in Nature, and less possible than a *Chimera* [a fire-breathing monster in Greek mythology, composed of a serpent, lion, and goat] or a *Centaure* [in Greek mythology, a being having the upper portion of a man and the body and legs of a horse]. For Length, Breadth and Thickness, take up the whole of *Space.* Nor can our Fansie imagine how there should be a Fourth Local Dimension beyond these Three.

Again, Wallis's logic here was clear: There was no point in even imagining a geometry that did not describe real space.

Opinions eventually started to change. Mathematicians of the eighteenth century were the first to consider time as a potential fourth dimension. In an article entitled "Dimension," published in 1754, the physicist Jean D'Alembert (1717–83) wrote:

> I stated above that it is impossible to conceive of more than three dimensions. A man of parts, of my acquaintance, holds that one may however look upon duration as a fourth dimen-

> sion, and that the product of time and solidity is in a way a product of four dimensions. This idea may be challenged but it seems to me to have some merit other than that of mere novelty.

The great mathematician Joseph Lagrange went even one step further, stating more assertively in 1797:

> Since a position of a point in space depends upon three rectangular coordinates these coordinates in the problems of mechanics are conceived as being functions of t [time]. Thus we may regard mechanics as a geometry of four dimensions, and mechanical analysis as an extension of geometrical analysis.

These bold ideas opened the door for extensions of mathematics that had previously been considered inconceivable—geometries in any number of dimensions—which totally ignored the question of whether they had any relation to physical space.

Kant may have been wrong in believing that our senses of spatial perception follow exclusively Euclidean molds, but there is no question that our perception operates most naturally and intuitively in no more than three dimensions. We can relatively easily imagine how our three-dimensional world would look in Plato's two-dimensional universe of shadows, but going beyond three to a higher number of dimensions truly requires a mathematician's imagination.

Some of the groundbreaking work in the treatment of *n-dimensional geometry*—geometry in an arbitrary number of dimensions—was carried out by Hermann Günther Grassmann (1809–77). Grassmann, one of twelve children, and himself the father of eleven, was a schoolteacher who never had any university mathematical training. During his lifetime, he received more recognition for his work in linguistics (in particular for his studies of Sanskrit and Gothic) than for his achievements in mathematics. One of his biographers wrote: "It seems to be Grassmann's fate to be rediscovered from time to time, each time as if he had been virtually forgotten since his death." Yet, Grassmann was responsible for the creation of an abstract science of "spaces," inside which the usual geometry was only a special case. Grassmann published his pioneering ideas (originating a branch of mathematics known as *linear algebra*) in 1844, in a book commonly known as the *Ausdehnungslehre* (meaning *Theory of Extension;* the full title read: *Linear Extension Theory: A New Branch of Mathematics*).

In the foreword to the book Grassmann wrote: "Geometry can in no way be viewed . . . as a branch of mathematics; instead, geometry relates to something already given in nature, namely, space. I also had realized that there must be a branch of mathematics which yields in a purely abstract way laws similar to those of geometry."

This was a radically new view of the nature of mathematics. To Grassmann, the traditional geometry—the heritage of the ancient Greeks—deals with physical space and therefore cannot be taken as a true branch of abstract mathematics. Mathematics to him was rather an abstract construct of the human brain that does not necessarily have any application to the real world.

It is fascinating to follow the seemingly trivial train of thought that set Grassmann on the road to his theory of geometric algebra. He started with the simple formula $AB + BC = AC$, which appears in any geometry book in the discussion of lengths of line segments (see figure 46a). Here, however, Grassmann noticed something interesting. He discovered that this formula remains valid irrespective of the order of the points A, B, C as long as one does not interpret AB, BC, and so on merely as lengths, but also assigns to them "direction," such that $BA = -AB$. For instance, if C lies between A and B (as in Figure 46b), then $AB = AC + CB$, but since $CB = -BC$, we find that $AB = AC - BC$ and the original formula $AB + BC = AC$ is recovered simply by adding BC to both sides.

This was quite interesting in itself, but Grassmann's extension contained even more surprises. Note that if we were dealing with algebra instead of geometry, then an expression such as AB usually would

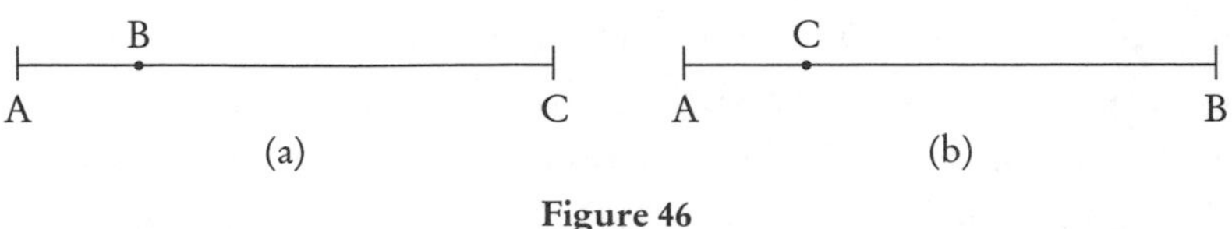

Figure 46

denote the product $A \times B$. In that case, Grassmann's suggestion of $BA = -AB$ violates one of the sacrosanct laws of arithmetic—that two quantities multiplied together produce the same result irrespective of the order in which the quantities are taken. Grassmann faced up squarely to this disturbing possibility and invented a new consistent algebra (known as *exterior algebra*) that allowed for several processes of multiplication and at the same time could handle geometry in any number of dimensions.

By the 1860s n-dimensional geometry was spreading like mushrooms after a rainstorm. Not only had Riemann's seminal lecture established spaces of any curvature and of arbitrary numbers of dimensions as a fundamental area of research, but other mathematicians, such as Arthur Cayley and James Sylvester in England, and Ludwig Schläfli in Switzerland, were adding their own original contributions to the field. Mathematicians started to feel that they were being freed from the restrictions that for centuries had tied mathematics only to the concepts of space and number. Those ties had historically been taken so seriously that even as late as the eighteenth century, the prolific Swiss mathematician Leonhard Euler (1707–83) expressed his view that "mathematics, in general, is the science of quantity; or, the science that investigates the means of measuring quantity." It was only in the nineteenth century that the winds of change started to blow.

First, the introduction of abstract geometric spaces and of the notion of infinity (in both geometry and the theory of sets) had blurred the meaning of "quantity" and of "measurement" beyond recognition. Second, the rapidly multiplying studies of mathematical abstractions helped to distance mathematics even further from physical reality, while breathing life and "existence" into the abstractions themselves.

Georg Cantor (1845–1918), the creator of *set theory*, characterized the newly found spirit of freedom of mathematics by the following "declaration of independence": "Mathematics is in its development entirely free and is only bound in the self-evident respect that its concepts must both be consistent with each other and also stand in exact relationships, ordered by definitions, to those concepts which have previously been introduced and are already at hand and established." To which algebraist Richard Dedekind (1831–1916) added six years later: "I consider the number concept entirely independent of the notions or intuitions of space and time . . . Numbers are free creations of the human mind." That is, both Cantor and Dedekind viewed mathematics as an abstract, conceptual investigation, constrained only by the requirement of consistency, with no obligations whatsoever toward either calculation or the language of physical reality. As Cantor has summarized it: "The *essence of mathematics* lies entirely in its *freedom*."

By the end of the nineteenth century most mathematicians accepted Cantor's and Dedekind's views on the freedom of mathematics. The objective of mathematics changed from being the search for truths about nature to the construction of abstract structures—systems of axioms—and the pursuit of all the logical consequences of those axioms.

One might have thought that this would put an end to all the agonizing over the question of whether mathematics was discovered or invented. If mathematics was nothing more than a game, albeit a complex one, played with arbitrarily invented rules, then clearly there was no point in believing in the reality of mathematical concepts, was there?

Surprisingly, the breaking away from physical reality infused some mathematicians with precisely the opposite sentiment. Rather than concluding that mathematics was a human invention, they returned to the original Platonic notion of mathematics as an independent world of truths, whose existence was as real as that of the physical universe. The attempts to connect mathematics with physics were treated by these "neo-Platonists" as dabbling in *applied* mathematics, as opposed to the *pure* mathematics that was supposed to be indifferent to anything physical. Here is how the French mathematician Charles Hermite (1822–1901) put it in a letter written to the Dutch mathematician Thomas Joannes Stieltjes (1856–94) on May 13, 1894: "My dear friend," he wrote,

> I feel very happy to find you inclined to transform yourself into a naturalist to observe the phenomena of the arithmetical

> world. Your doctrine is the same as mine; I believe that numbers and the functions of analysis are not arbitrary products of our mind; I think that they exist outside of us with the same necessary characteristics as the things of objective reality, and that we encounter them or discover them, and study them, just as the physicists, the chemists and the zoologists.

The English mathematician G. H. Hardy, himself a practitioner of pure mathematics, was one of the most outspoken modern Platonists. In an eloquent address to the British Association for the Advancement of Science on September 7, 1922, he pronounced:

> Mathematicians have constructed a very large number of different systems of geometry. Euclidean or non-Euclidean, of one, two, three, or any number of dimensions. All these systems are of complete and equal validity. They embody the results of mathematicians' observations of their reality, a reality far more intense and far more rigid than the dubious and elusive reality of physics . . . The function of a mathematician, then, is simply to observe the facts about his own hard and intricate system of reality, that astonishingly beautiful complex of logical relations which forms the subject matter of his science, as if he were an explorer looking at a distant range of mountains, and to record the results of his observations in a series of maps, each of which is a branch of pure mathematics.

Clearly, even with the contemporary evidence pointing to the arbitrary nature of mathematics, the die-hard Platonists were not about to lay down their arms. Quite the contrary, they found the opportunity to delve into, in Hardy's words, "their reality," even more exciting than to continue to explore the ties to physical reality. Irrespective, however, of the opinions on the metaphysical reality of mathematics, one thing was becoming obvious. Even with the seemingly unbridled freedom of mathematics, one constraint remained unchanging and unshakable—that of logical consistency. Mathematicians and philosophers were becoming more aware than ever that the umbilical cord between mathematics and logic could not be cut. This gave birth to another idea: Could all of mathematics be built on a single logical foundation? And if it could, was that the secret of its effectiveness? Or conversely, could mathematical methods be used in the study of reasoning in general? In which case, mathematics would become not just the language of nature, but also the language of human thought.

CHAPTER 7

LOGICIANS: THINKING ABOUT REASONING

The sign outside a barber shop in one village reads: "I shave all and only those men in the village who do not shave themselves." Sounds perfectly reasonable, right? Clearly, the men who shave themselves do not need the services of the barber, and it is only natural for the barber to shave everyone else. But, ask yourself, who shaves the barber? If he shaves himself, then according to the sign he should be one of those he does not shave. On the other hand, if he does not shave himself, then again according to the sign he should be one of those he does shave! So does he or doesn't he? Much lesser questions have historically resulted in serious family feuds. This paradox was introduced by Bertrand Russell (1872–1970), one of the most prominent logicians and philosophers of the twentieth century, simply to demonstrate that human logical intuition is fallible. Paradoxes or *antinomies* reflect situations in which apparently acceptable premises lead to unacceptable conclusions. In the example above, the village barber both shaves and doesn't shave himself. Can this particular paradox be resolved? One possible resolution to the paradox, strictly as stated above, is simple: The barber is a woman! On the other hand, were we told in advance that the barber had to be a man, then the absurd conclusion would have been the result of accepting the premise in the first place. In other words, such a barber simply cannot exist. But what does any of this have to do with mathematics? As it turns out, mathematics and logic are intimately related. Here is how Russell himself described the linkage:

> Mathematics and logic, historically speaking, have been entirely distinct studies. Mathematics has been connected with science, logic with Greek. But both have developed in modern times: logic has become more mathematical and mathematics has become more logical. The consequence is that it has now [in 1919] become wholly impossible to draw a line between the two; in fact the two are one. They differ as boy and man: logic is the youth of mathematics and mathematics is the manhood of logic.

Russell holds here that, largely, *mathematics can be reduced to logic.* In other words, that the basic concepts of mathematics, even objects such as numbers, can in fact be defined in terms of the fundamental laws of reasoning. Furthermore, Russell would later argue that one can use those definitions in conjunction with logical principles to give birth to the theorems of mathematics.

Originally, this view of the nature of mathematics (known as *logicism*) had received the blessing of both those who regarded mathematics as nothing but a human-invented, elaborate game (the *formalists*), and the troubled Platonists. The former were initially happy to see a collection of seemingly unrelated "games" coalesce into one "mother of all games." The latter saw a ray of hope in the idea that the whole of mathematics could have stemmed from one indubitable source. In the Platonists' eyes, this enhanced the probability of a single metaphysical origin. Needless to say, a single root of mathematics could have also helped, in principle at least, to identify the cause for its powers.

For completeness, I should note that there was one school of thought—*intuitionism*—that was vehemently opposed to both logicism and formalism. The torch-bearer of this school was the rather fanatical Dutch mathematician Luitzen E. J. Brouwer (1881–1966). Brouwer believed that the natural numbers derive from a human intuition of time and of discrete moments in our experience. To him, there was no question that mathematics was a result of human thought, and he therefore saw no need for universal logical laws of the type that Russell envisioned. Brouwer did go much further, however, and declared that the only meaningful mathematical entities were those that could be explicitly constructed on the basis of the natural numbers, using a finite number of steps. Consequently, he rejected large parts of mathematics for which constructive proofs were not possible. Another logical concept denied by Brouwer was the *principle of the excluded middle*—the stipulation that any statement is either true or false. Instead, he allowed for statements to linger in a third limbo state in which they were "undecided." These, and a few other intuitionist limiting constraints, somewhat marginalized this school of thought. Nevertheless, intuitionist ideas did anticipate some of the findings of cognitive scientists concerning the question of how humans actually acquire mathematical knowledge (a topic to be discussed in chapter 9), and they also informed the discussions of some modern philosophers of mathematics (such as Michael Dummett). Dummett's approach is basically linguistic, stating forcefully that "the meaning of a mathematical statement determines and is exhaustively determined by its *use*."

But how did such a close partnership between mathematics and logic develop? And was the logicist program at all viable? Let me briefly review a few of the milestones of the last four centuries.

Logic and Mathematics

Traditionally, logic dealt with the relationships between concepts and propositions and with the processes by which valid inferences could be distilled from those relationships. As a simple example, inferences of the general form "every *X* is a *Y*; some *Z*'s are *X*'s; therefore some *Z*'s are *Y*'s" are constructed so as to automatically ensure the truth of the conclusion, as long as the premises are true. For instance, "every biographer is an author; some politicians are biographers; therefore some politicians are authors" produces a true conclusion. On the other hand, inferences of the general form "every *X* is a *Y*; some *Z*'s are *Y*'s; therefore, some *Z*'s are *X*'s" are not valid, since one can find examples where in spite of the premises being true, the conclusion is false. For example: "every man is a mammal; some horned animals are mammals; therefore, some horned animals are men."

As long as some rules are being followed, the validity of an argument does not depend on the subjects of the statements. For instance:

> Either the butler murdered the millionaire or his daughter
> killed him;
> His daughter did not kill him;
> Therefore the butler murdered him.

produces a valid deduction. The soundness of this argument does not rely at all on our opinion of the butler or on the relationship between the millionaire and his daughter. The validity here is ensured by the fact that propositions of the general form "if either p or q, and not q, then p" yield logical truth.

You may have noticed that in the first two examples X, Y, and Z play roles very similar to those of the variables in mathematical equations—they mark the place where expressions can be inserted, in the same way that numerical values are inserted for variables in algebra. Similarly, the truth in the inference "if either p or q, and not q, then p" is reminiscent of the axioms in Euclid's geometry. Still, nearly two millennia of contemplation of logic had to pass before mathematicians took this analogy to heart.

The first person to have attempted to combine the two disciplines of logic and mathematics into one "universal mathematics" was the German mathematician and rationalist philosopher Gottfried Wilhelm Leibniz (1646–1716). Leibniz, whose formal training was in law, did most of his work on mathematics, physics, and philosophy in his spare time. During his lifetime, he was best known for formulating independently of (and almost simultaneously with) Newton the foundations of calculus (and for the ensuing bitter dispute between them on priority). In an essay conceived almost entirely at age sixteen, Leibniz envisaged a universal language of reasoning, or *characteristica universalis,* which he regarded as the ultimate thinking tool. His plan was to represent simple notions and ideas by symbols, more complex ones by appropriate combinations of those basic signs. Leibniz hoped to be able to literally compute the truth of any statement, in any scientific discipline, by mere algebraic operations. He prophesied that with the proper logical calculus, debates in philosophy would be resolved by calculation. Unfortunately, Leibniz did not get very far in actually developing his algebra of logic. In addition to the general principle of an "alphabet of thought," his two main contributions have been a clear statement about when we should view two things as equal and the somewhat obvious recognition that no statement can be true and false at the same time. Consequently, even though Leibniz's ideas were scintillating, they went almost entirely unnoticed.

Logic became more in vogue again in the middle of the nineteenth century, and the sudden surge in interest produced important works, first by Augustus De Morgan (1806–71) and later by George Boole (1815–64), Gottlob Frege (1848–1925), and Giuseppe Peano (1858–1932).

De Morgan was an incredibly prolific writer who published literally thousands of articles and books on a variety of topics in mathematics, the history of mathematics, and philosophy. His more unusual work included an almanac of full moons (covering millennia) and a compendium of eccentric mathematics. When asked once about his age he replied: "I was x years old in the year x^2." You can check that the only number that, when squared, gives a number between 1806 and 1871 (the years of De Morgan's birth and death) is 43. De Morgan's most original contributions were still probably in the field of logic, where he both considerably expanded the scope of Aristotle's syllogisms and rehearsed an algebraic approach to reasoning. De Morgan looked at logic with the eyes of an algebraist and at algebra with the eyes of a logician. In one of his articles he described this visionary perspective: "It is to algebra that we must look for the most habitual use of logical forms . . . the algebraist was living in the higher atmosphere of syllogism, the unceasing composition of relation, before it was admitted that such an atmosphere existed."

One of De Morgan's most important contributions to logic is known as *quantification of the predicate.* This is a somewhat bombastic name for what one might view as a surprising oversight on the part of the logicians of the classical period. Aristotelians correctly realized that from premises such as "some Z's are X's" and "some Z's are Y's" no conclusion of necessity can be reached about the relation between the *X*'s and the *Y*'s. For instance, the phrases "some people eat bread" and "some people eat apples" permit no decisive conclusions about the relation between the apple eaters and the bread eaters. Until the nineteenth century, logicians also assumed that for any relation between the *X*'s and the *Y*'s to follow of necessity, the middle term ("Z" above) must be "universal" in one of the premises. That is, the phrase must include "all Z's." De Morgan showed this assumption to be wrong. In his book *Formal Logic* (published in 1847), he pointed out that from premises such as "most *Z*'s are *X*'s" and "most *Z*'s are *Y*'s" it necessarily follows that "some *X*'s are *Y*'s." For instance, the phrases "most people eat bread" and "most people eat apples" inevitably imply that "some people eat both bread and apples." De Morgan went even further and put his new syllogism in precise quantitative form. Imagine that the total number of *Z*'s is z, the number of *Z*'s that are also *X*'s is x, and the number of *Z*'s that are also *Y*'s is y. In the above example, there could be 100 people in total ($z = 100$), of which 57 eat bread ($x = 57$) and 69 eat apples ($y = 69$). Then, De Morgan noticed, there must be at least ($x + y - z$) *X*'s that are also *Y*'s. At least 26 people (obtained from $57 + 69 - 100 = 26$) eat both bread and apples.

Unfortunately, this clever method of quantifying the predicate dragged De Morgan into an unpleasant public dispute. The Scottish philosopher William Hamilton (1788–1856)—not to be confused with the Irish mathematician William Rowan Hamilton—accused De Morgan of plagiarism, because Hamilton had published somewhat related (but much less accurate) ideas a few years before De Morgan. Hamilton's attack was not at all surprising, given his general attitude toward mathematics and mathematicians. He once said: "An excessive study of mathematics absolutely incapacitates the mind for those intellectual energies which philosophy and life require." The flurry of acrimonious letters that followed Hamilton's accusation produced one positive, if totally unintended, result: It guided algebraist George Boole to logic. Boole later recounted in *The Mathematical Analysis of Logic:*

> In the spring of the present year my attention was directed to the question then moved between Sir W. Hamilton and Professor

> De Morgan; and I was induced by the interest which it inspired, to resume the almost-forgotten thread of former inquiries. It appeared to me that, although Logic might be viewed with reference to the idea of quantity, it had also another and a deeper system of relations. If it was lawful to regard it from *without,* as connecting itself through the medium of Number with the intuitions of Space and Time, it was lawful also to regard it from *within,* as based upon facts of another order which have their abode in the constitution of the Mind.

These humble words describe the initiation of what was to become a seminal effort in symbolic logic.

Laws of Thought

George Boole (figure 47) was born on November 2, 1815, in the industrial town of Lincoln, England. His father, John Boole, a shoemaker in Lincoln, showed great interest in mathematics and was skilled in the construction of a variety of optical instruments. Boole's mother, Mary Ann Joyce, was a lady's maid. With the father's heart not quite

Figure 47

in his formal business, the family was not well off. George attended a commercial school until age seven, and then a primary school, where his teacher was one John Walter Reeves. As a boy, Boole was mainly interested in Latin, in which he received instruction from a bookseller, and in Greek, which he learned by himself. At age fourteen he managed to translate a poem by the first century BC Greek poet Meleager. George's proud father published the translation in the *Lincoln Herald*—an act that provoked an article expressing disbelief from a local schoolmaster. The poverty at home forced George Boole to start working as an assistant teacher at age sixteen. During the following years, he devoted his free time to the study of French, Italian, and German. The knowledge of these modern languages proved useful, since it allowed him to turn his attention to the great works of mathematicians such as Sylvestre Lacroix, Laplace, Lagrange, Jacobi, and others. Even then, however, he was still unable to take regular courses in mathematics, and he continued to study on his own, while at the same time helping to support his parents and siblings through his teaching job. Nevertheless, the mathematical talents of this autodidact started to show, and he began publishing articles in the *Cambridge Mathematical Journal.*

In 1842, Boole started to correspond regularly with De Morgan, to whom he was sending his mathematical papers for comments. Because of his growing reputation as an original mathematician, and backed by a strong recommendation from De Morgan, Boole was offered the position of professor of mathematics at Queen's College, in Cork, Ireland, in 1849. He continued to teach there for the rest of his life. In 1855 Boole married Mary Everest (after whose uncle, the surveyor George Everest, the mountain was named), who was seventeen years his junior, and the couple had five daughters. Boole died prematurely at age forty-nine. On a cold winter day in 1864, he got drenched on his way to the college, but he insisted on giving his lectures even though his clothes were soaking wet. At home, his wife may have worsened his condition by pouring buckets of water onto the bed, following a superstition that the cure should somehow replicate the cause of the illness. Boole developed pneumonia and died on December 8, 1864. Bertrand Russell did not hide his admiration for this self-taught individual: "Pure mathematics was discovered by Boole, in a work which he called *The Laws of Thought* (1854) . . . His book was in fact concerned with formal logic, and this is the same thing as mathematics." Remarkably for that time, both Mary Boole (1832–1916) and each of the five Boole daughters achieved considerable fame in fields ranging from education to chemistry.

Boole published *The Mathematical Analysis of Logic* in 1847 and *The Laws of Thought* in 1854 (the full title of the latter read: *An Investigation of the Laws of Thought, on Which Are Founded the Mathematical Theories of Logic and Probabilities*). These were genuine masterworks—the first to take the parallelism between logical and arithmetic operations a giant step forward. Boole literally transformed logic into a type of algebra (which came to be called *Boolean algebra*) and extended the analysis of logic even to probabilistic reasoning. In Boole's words:

> The design of the following treatise [*The Laws of Thought*] is to investigate the fundamental laws of those operations of the mind by which reasoning is performed; to give expression to them in the symbolical language of a Calculus, and upon this foundation to establish the science of Logic and construct its method; to make that method itself the basis of a general method for the application of the mathematical doctrine of Probabilities; and finally to collect from the various elements of truth brought to view in the course of these inquiries some probable intimations concerning the nature and constitution of the human mind.

Boole's calculus could be interpreted either as applying to relations among *classes* (collections of objects or members) or within the logic of *propositions*. For instance, if x and y were classes, then a relation such as $x = y$ meant that the two classes had precisely the same members, even if the classes were defined differently. As an example, if all the children in a certain school are shorter than seven feet, then the two classes defined as x = "all the children in the school" and y = "all the children in the school that are shorter than seven feet" are equal. If x and y represented propositions, then $x = y$ meant that the two propositions were equivalent (that one was true if and only if the other was also true). For example, the propositions x = "John Barrymore was Ethel Barrymore's brother" and y = "Ethel Barrymore was John Barrymore's sister" are equal. The symbol "$x \cdot y$" represented the common part of the two classes x and y (those members belonging to both x and y), or the *conjunction* of the propositions x and y (i.e., "x *and* y"). For instance, if x was the class of all village idiots and y was the class of all things with black hair, then $x \cdot y$ was the class of all black-haired village idiots. For propositions x and y, the conjunction $x \cdot y$ (or the word "and") meant that both propositions had to hold. For example, when the Motor Vehicle Administration says that "you must pass a peripheral vision test and a driving test," this means that both requirements must be met. For Boole, for two classes having no members in common, the symbol "$x + y$" represented the class consisting of both the members of x and the members of y. In the case of propositions, "$x + y$" corresponded to "either x or y but not both." For instance, if x is the proposition "pegs are square" and y is "pegs are round," then $x + y$ is "pegs are either square or round." Similarly, "$x - y$" represented the class of those members of x that were not members of y, or the proposition "x but not y." Boole denoted the universal class (containing all possible members under discussion) by 1 and the empty or null class (having no members whatsoever) by 0. Note that the null class (or set) is definitely not the same as the number 0—the latter is simply the number of members in the null class. Note also that the null class is not the same as nothing, because a class with nothing in it is still a class. For instance, if all the newspapers in Albania are written in Albanian, then the class of all Albanian-language newspapers in Albania would be denoted by 1 in Boole's notation, while the class of all Spanish-language newspapers in Albania would be denoted by 0. For propositions, 1 represented the standard *true* (e.g., humans are mortal) and 0 the standard *false* (e.g., humans are immortal) propositions, respectively.

With these conventions, Boole was able to formulate a set of axioms defining an algebra of logic. For instance, you can check that using the above definitions, the obviously true proposition "everything is either x or not x" could be written in Boole's algebra as $x + (1 - x) = 1$, which also holds in ordinary algebra. Similarly, the statement that the common part between any class and the empty class is an empty class was represented by $0 \cdot x = 0$, which also meant that the conjunction of any proposition with a false proposition is false. For instance, the conjunction "sugar is sweet and humans are immortal" produces a false proposition in spite of the fact that the first part is true. Note that again, this "equality" in Boolean algebra holds true also with normal algebraic numbers.

To show the power of his methods, Boole attempted to use his logical symbols for everything he deemed important. For instance, he even analyzed the arguments of the philosophers Samuel Clarke and Baruch Spinoza for the existence and attributes of God. His conclusion, however, was rather pessimistic: "It is not possible, I think, to rise from the perusal of the arguments of Clarke and Spinoza without a deep conviction of the futility of all endeavors to establish, entirely *a priori,* the existence of an Infinite Being, His attributes, and His relation to the universe." In spite of the soundness of Boole's conclusion, apparently not everybody was convinced of the futility of such endeavors, since updated versions of ontological arguments for God's existence continue to emerge even today.

Overall, Boole managed to mathematically tame the logical connectives *and, or, if . . . then,* and *not,* which are currently at the very core of computer operations and various switching circuits. Consequently, he is regarded by many as one of the "prophets" who brought about the digital age. Still, due to its pioneering nature, Boole's algebra was not perfect. First, Boole made his writings somewhat ambiguous and difficult to comprehend by using a notation that was too close to that of ordinary algebra. Second, Boole confused the distinction between propositions (e.g., "Aristotle is mortal"), propositional functions or predicates (e.g., "x is mortal"), and quantified statements (e.g., "for all x, x is mortal"). Finally, Frege and Russell were later to claim that algebra stems from logic. One could argue, therefore, that it made more sense to construct algebra on the basis of logic rather than the other way around.

There was another aspect of Boole's work, however, that was

about to become very fruitful. This was the realization of how closely related logic and the concept of *classes* or *sets* were. Recall that Boole's algebra applied equally well to classes and to logical propositions. Indeed, when all the members of one set *X* are also members of set *Y* (*X* is a *subset* of *Y*), this fact can be expressed as a *logical implication* of the form "if *X* then *Y*." For instance, the fact that the set of all horses is a subset of the set of all four-legged animals can be rewritten as the logical statement "If *X* is a horse then it is a four-legged animal."

Boole's algebra of logic was subsequently expanded and improved upon by a number of researchers, but the person who fully exploited the similarity between sets and logic, and who took the entire concept to a whole new level, was Gottlob Frege (figure 48).

Friedrich Ludwig Gottlob Frege was born at Wismar, Germany, where both his father and his mother were, at different times, the principals at a girls' high school. He studied mathematics, physics,

Figure 48

chemistry, and philosophy, first at the University of Jena and then for an additional two years at the University of Göttingen. After completing his education, he started lecturing at Jena in 1874, and he continued to teach mathematics there throughout his entire professional career. In spite of a heavy teaching load, Frege managed to publish his first revolutionary work in logic in 1879. The publication was entitled *Concept-Script, A Formal Language for Pure Thought Modeled on that of Arithmetic* (it is commonly known as the *Begriffsschrift*). In this work, Frege developed an original, logical language, which he later amplified in his two-volume *Grundgesetze der Arithmetic* (*Basic Laws of Arithmetic*). Frege's plan in logic was on one hand very focused, but on the other extraordinarily ambitious. While he primarily concentrated on arithmetic, he wanted to show that even such familiar concepts as the natural numbers, 1, 2, 3, . . . , could be reduced to logical constructs. Consequently, Frege believed that one could prove all the truths of arithmetic from a few axioms in logic. In other words, according to Frege, even statements such as 1 + 1 = 2 were not *empirical* truths, based on observation, but rather they could be derived from a set of logical axioms. Frege's *Begriffsschrift* has been so influential that the contemporary logician Willard Van Orman Quine (1908–2000) once wrote: "Logic is an old subject, and since 1879 it has been a great one."

Central to Frege's philosophy was the assertion that truth is independent of human judgment. In his *Basic Laws of Arithmetic* he writes: "Being true is different from being taken to be true, whether by one or many or everybody, and in no case is it to be reduced to it. There is no contradiction in something's being true which everybody takes to be false. I understand by 'laws of logic' not psychological laws of takings-to-be-true, but laws of truth . . . they [the laws of truth] are boundary stones set in an eternal foundation, which our thought can overflow, but never displace."

Frege's logical axioms generally take the form "for all . . . if . . . then . . ." For instance, one of the axioms reads: "for all *p,* if not-(not-*p*) then *p*." This axiom basically states that if a proposition that is contradictory to the one under discussion is false, then the proposition is true. For instance, if it is not true that you do not have to stop your car at a stop sign, then you definitely do have to stop at a stop sign. To actually develop a logical "language," Frege supplemented the set of axioms with an important new feature. He replaced the traditional subject/predicate style of classical logic by concepts borrowed from the mathematical theory of functions. Let me briefly explain. When one writes in mathematics expressions such as: $f(x) = 3x + 1$, this means that f is a function of the variable x and that the value of the function can be obtained by multiplying the value of the variable by three and then adding one. Frege defined what he called *concepts* as functions. For example, suppose you want to discuss the concept "eats meat." This concept would be denoted symbolically by a function "$F(x)$," and the value of this function would be "true" if x = lion, and "false" if x = deer. Similarly, with respect to numbers, the concept (function) "being smaller than 7" would map every number equal to or larger than 7 to "false" and all numbers smaller than 7 to "true." Frege referred to objects for which a certain concept gave the value of "true" as "falling under" that concept.

As I noted above, Frege firmly believed that every proposition concerning the natural numbers was knowable and derivable solely from logical definitions and laws. Accordingly, he started his exposition of the subject of natural numbers without requiring any prior understanding of the notion of "number." For instance, in Frege's logical language, two concepts are *equinumerous* (that is, they have the same number associated with them) if there is a one-to-one correspondence between the objects "falling under" one concept and the objects "falling under" the other. That is, garbage can lids are equinumerous with the garbage cans themselves (if every can has a lid), and this definition does not require any mention of numbers. Frege then introduced an ingenious logical definition of the number 0. Imagine a concept F defined by "not identical to itself." Since every object has to be identical to itself, no objects fall under F. In other words, for any object x, $F(x)$ = false. Frege defined the common number zero as being the "number of the concept F." He then went on to define all the natural numbers in terms of entities he called *extensions.* The extension of a concept was the class of all the objects that fall under that concept. While this definition may not be the easiest to digest for the nonlogician, it is really quite simple. The extension of the concept "woman," for instance, was the class of all women. Note that the extension of "woman" is not in itself a woman.

You may wonder how this abstract logical definition helped to define, say, the number 4. According to Frege, the number 4 was the extension (or class) of all the concepts that have four objects falling under them. So, the concept "being a leg of a particular dog named Snoopy" belongs to that class (and therefore to the number 4), as does the concept "being a grandparent of Gottlob Frege."

Frege's program was extraordinarily impressive, but it also suffered from some serious drawbacks. On one hand, the idea of using concepts—the bread and butter of thinking—to construct arithmetic, was pure genius. On the other, Frege did not detect some crucial inconsistencies in his formalism. In particular, one of his axioms—known as *Basic Law V*—proved to lead to a contradiction and was therefore fatally flawed.

The law itself stated innocently enough that the extension of a concept F is identical to the extension of concept G if and only if F and G have the same objects under them. But the bomb was dropped on June 16, 1902, when Bertrand Russell (figure 49) wrote a letter to

Figure 49

Frege, pointing out to him a certain paradox that showed Basic Law V to be inconsistent. As fate would have it, Russell's letter arrived just as the second volume of Frege's *Basic Laws of Arithmetic* was going to press. The shocked Frege hastened to add to the manuscript the frank admission: "A scientist can hardly meet with anything more undesirable than to have the foundations give way just as the work is finished. I was put in this position by a letter from Mr. Bertrand Russell when the work was nearly through the press." To Russell himself, Frege graciously wrote: "Your discovery of the contradiction caused me the greatest surprise and, I would almost say, consternation, since it has shaken the basis on which I intended to build arithmetic."

The fact that one paradox could have such a devastating effect on an entire program aimed at creating the bedrock of mathematics may sound surprising at first, but as Harvard University logician W. V. O. Quine once noted: "More than once in history the discovery of paradox has been the occasion for major reconstruction at the foundation of thought." Russell's paradox provided for precisely such an occasion.

Russell's Paradox

The person who essentially single-handedly founded the theory of sets was the German mathematician Georg Cantor. Sets, or classes, quickly proved to be so fundamental and so intertwined with logic that any attempt to build mathematics on the foundation of logic necessarily implied that one was building it on the axiomatic foundation of set theory.

A class or a set is simply a collection of objects. The objects don't have to be related in any way. You can speak of one class containing all of the following items: the soap operas that aired in 2003, Napoleon's white horse, and the concept of true love. The elements that belong to a certain class are called *members* of that class.

Most classes of objects you are likely to come up with are not members of themselves. For instance, the class of all snowflakes is not in itself a snowflake; the class of all antique watches is not an antique watch, and so on. But some classes actually are members of themselves. For example, the class of "everything that is not an antique watch" is a member of itself, since this class is definitely not an antique watch. Similarly, the class of all classes is a member of itself since obviously it is a class. How about, however, the class of "all of those classes that are not members of themselves"? Let's call that class *R*. Is *R* a member of itself (of *R*) or not? Clearly *R* cannot belong to *R*, because if it did, it would violate the definition of the *R* membership. But if *R* does not belong to itself, then according to the definition it must be a member of *R*. Similar to the situation with the village barber, we therefore find that the class *R* both belongs and does not belong to *R*, which is a logical contradiction. This was the paradox that Russell sent to Frege. Since this antinomy undermined the entire process by which classes or sets could be determined, the blow to Frege's program was deadly. While Frege did make some desperate attempts to remedy his axiom system, he was unsuccessful. The conclusion appeared to be disastrous—rather than being more solid than mathematics, formal logic appeared to be more vulnerable to paralyzing inconsistencies.

Around the same time that Frege was developing his logicist program, the Italian mathematician and logician Giuseppe Peano was attempting a somewhat different approach. Peano wanted to base arithmetic on an axiomatic foundation. Consequently, his starting point was the formulation of a concise and simple set of axioms. For instance, his first three axioms read:

1. Zero is a number.
2. The successor to any number is also a number.
3. No two numbers have the same successor.

The problem was that while Peano's axiomatic system could indeed reproduce the known laws of arithmetic (when additional definitions had been introduced), there was nothing about it that uniquely identified the natural numbers.

The next step was taken by Bertrand Russell. Russell maintained that Frege's original idea—that of deriving arithmetic from logic—was still the right way to go. In response to this tall order, Russell produced, together with Alfred North Whitehead (figure 50), an incredible logical masterpiece— the landmark three-volume *Principia*

Figure 50

Mathematica. With the possible exception of Aristotle's *Organon*, this has probably been the most influential work in the history of logic (figure 51 shows the title page of the first edition).

In the *Principia*, Russell and Whitehead defended the view that mathematics was basically an elaboration of the laws of logic, with no clear demarcation between them. To achieve a self-consistent description, however, they still had to somehow bring the antino mies or paradoxes (additional ones to Russell's paradox had been discovered) under control. This required some skillful logical juggling. Russell argued that those paradoxes arose only because of a "vicious circle" in which one was defining entities in terms of a class of objects that in itself contained the defined entity. In Russell's words: "If I say 'Napoleon had all the qualities that make a great general,' I must define 'qualities' in such a way that it will not include what I am now saying, i.e. 'having all the qualities that make a great general' must not be itself a quality in the sense supposed."

To avoid the paradox, Russell proposed a *theory of types*, in which a class (or set) belongs to a higher logical type than that to which its members belong. For instance, all the individual players of the Dallas Cowboys football team would be of type 0. The Dallas Cow-

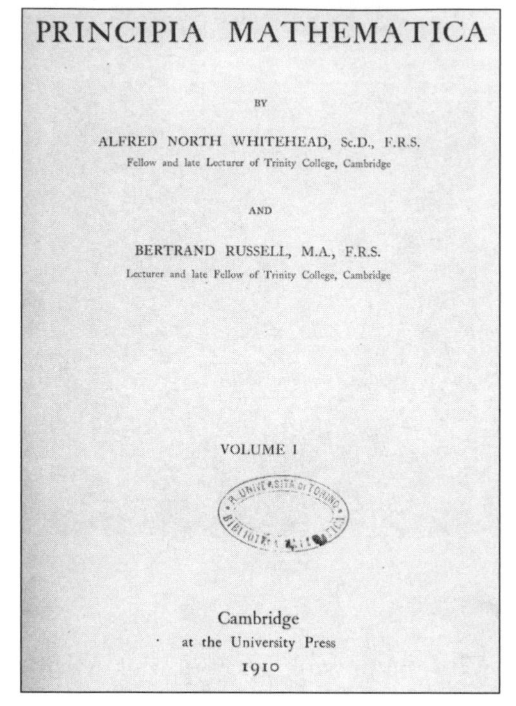
PRINCIPIA MATHEMATICA

BY

ALFRED NORTH WHITEHEAD, Sc.D., F.R.S.
Fellow and late Lecturer of Trinity College, Cambridge

AND

BERTRAND RUSSELL, M.A., F.R.S.
Lecturer and late Fellow of Trinity College, Cambridge

VOLUME I

Cambridge
at the University Press
1910

Figure 51

boys team itself, which is a class of players, would be of type 1. The National Football League, which is a class of teams, would be of type 2; a collection of leagues (if one existed) would be of type 3, and so on. In this scheme, the mere notion of "a class that is a member of itself" is neither true nor false, but simply meaningless. Consequently, paradoxes of the kind of Russell's paradox are never encountered.

There is no question that the *Principia* was a monumental achievement in logic, but it could hardly be regarded as the long-sought-for foundation of mathematics. Russell's theory of types was viewed by many as a somewhat artificial remedy to the problem of paradoxes—one that, in addition, produced disturbingly complex ramifications. For instance, rational numbers (e.g., simple fractions) turned out to be of a higher type than the natural numbers. To avoid some of these complications, Russell and Whitehead introduced an additional axiom, known as the axiom of reducibility, which in itself generated serious controversy and mistrust.

More elegant ways to eliminate the paradoxes were eventually suggested by the mathematicians Ernst Zermelo and Abraham Fraenkel. They, in fact, managed to self-consistently axiomatize set theory and to reproduce most of the set-theoretical results. This seemed, on the face of it, to be at least a partial fulfillment of the Platonists' dream. If set theory and logic were truly two faces of the same coin, then a solid foundation of set theory implied a solid foundation of logic. If, in addition, much of mathematics indeed followed from logic, then this gave mathematics some sort of objective certainty, which could also perhaps be harnessed to explain the effectiveness of mathematics. Unfortunately, the Platonists couldn't celebrate for very long, because they were about to be hit by a bad case of déjà vu.

The Non-Euclidean Crisis All Over Again?

In 1908, the German mathematician Ernst Zermelo (1871–1953) followed a path very similar to that originally paved by Euclid around 300 BC. Euclid formulated a few unproved but supposedly self-evident postulates about points and lines and then constructed geometry on the basis of those axioms. Zermelo—who discovered Russell's paradox independently as early as 1900—proposed a way to build set theory on a corresponding axiomatic foundation. Russell's paradox was bypassed in this theory by a careful choice of construction principles that eliminated contradictory ideas such as "the set of all sets." Zermelo's scheme was further augmented in 1922 by the Israeli mathematician Abraham Fraenkel (1891–1965) to form what has become known as the Zermelo-Fraenkel set theory (other important changes were added by John von Neumann in 1925). Things would have been nearly perfect (consistency was yet to be demonstrated) were it not for some nagging suspicions. There was one axiom—the *axiom of choice*—that just like Euclid's famous "fifth" was causing mathematicians serious heartburn. Put simply, the axiom of choice states: If X is a collection (set) of nonempty sets, then we can choose a single member from each and every set in X to form a new set Y. You can easily check that this statement is true if the collection X is not infinite. For instance, if we have one hundred boxes, each one containing at least one marble, we can easily choose one marble from each box to form a new set *Y* that contains one hundred marbles. In such a case, we do not need a special axiom; we can actually prove that a choice is possible. The statement is true even for infinite collections *X*, as long as we can precisely specify how we make the choice. Imagine, for instance, an infinite collection of nonempty sets of natural numbers. The members of this collection might be sets such as {2, 6, 7}, {1, 0}, {346, 5, 11, 1257}, {all the natural numbers between 381 and 10,457}, and so on. In every set of natural numbers, there is always one member that is the smallest. Our choice could therefore be uniquely described this way: "From each set we choose the smallest element." In this case again the need for the axiom of choice can be dodged. The problem arises for infinite collections in those instances in which we cannot define the choice. Under such circumstances the choice process never ends, and the existence of a set consisting of precisely one element from each of the members of the collection *X* becomes a matter of faith.

From its inception, the axiom of choice generated considerable controversy among mathematicians. The fact that the axiom asserts the existence of certain mathematical objects (e.g., choices), without actually providing any tangible example of one, has drawn fire, especially from adherents to the school of thought known as *constructivism* (which was philosophically related to *intuitionism*). The constructivists argued that anything that exists should also be explicitly constructible. Other mathematicians also tended to avoid the axiom of choice and only used the other axioms in the Zermelo-Fraenkel set theory.

Due to the perceived drawbacks of the axiom of choice, mathematicians started to wonder whether the axiom could either be proved using the other axioms or refuted by them. The history of Euclid's fifth axiom was literally repeating itself. A partial answer was finally given in the late 1930s. Kurt Gödel (1906–78), one of the most influential logicians of all time, proved that the axiom of choice and another famous conjecture due to the founder of set theory, Georg Cantor, known as the *continuum hypothesis*, were both consistent with the other Zermelo-Fraenkel axioms. That is, neither of the two hypotheses could be refuted using the other standard set theory axioms. Additional proofs in 1963 by the American mathematician Paul Cohen (1934–2007, who sadly passed away during the time I was writing this book) established the complete independence of the axiom of choice and the continuum hypothesis. In other words, the axiom of choice can neither be proved nor refuted from the other axioms of set theory. Similarly, the continuum hypothesis can neither be proved nor refuted from the same collection of axioms, even if one includes the axiom of choice.

This development had dramatic philosophical consequences. As in the case of the non-Euclidean geometries in the nineteenth century, there wasn't just one definitive set theory, but rather at least four! One could make different assumptions about infinite sets and end up with mutually exclusive set theories. For instance, one could assume that both the axiom of choice and the continuum hypothesis hold true and obtain one version, or that both do not hold, and obtain an entirely different theory. Similarly, assuming the validity of one of the two axioms and the negation of the other would have led to yet two other set theories.

This was the non-Euclidean crisis revisited, only worse. The fundamental role of set theory as the potential basis for the whole of mathematics made the problem for the Platonists much more acute. If indeed one could formulate many set theories simply by choosing a different collection of axioms, didn't this argue for mathematics being nothing but a human invention? The formalists' victory looked virtually assured.

An Incomplete Truth

While Frege was very much concerned with the meaning of axioms, the main proponent of formalism, the great German mathematician David Hilbert (figure 52), advocated complete avoidance of any interpretation of mathematical formulae. Hilbert was not interested in questions such as whether mathematics could be derived from logical notions. Rather, to him, mathematics proper consisted simply of a collection of meaningless formulae—structured patterns composed of arbitrary symbols. The job of guaranteeing the foundations of math-

Figure 52

ematics was assigned by Hilbert to a new discipline, one he referred to as "metamathematics." That is, metamathematics was concerned with using the very methods of mathematical analysis to prove that the entire process invoked by the formal system, of deriving theorems from axioms by following strict rules of inference, was consistent. Put differently, Hilbert thought that he could prove mathematically that mathematics works. In his words:

> My investigations in the new grounding of mathematics have as their goal nothing less than this: to eliminate, once and for all, the general doubt about the reliability of mathematical inference . . . Everything that previously made up mathematics is to be rigorously formalized, so that mathematics proper or mathematics in the strict sense becomes a stock of formulae . . . In addition to this formalized mathematics proper, we have a mathematics that is to some extent new: a metamathematics that is necessary for securing mathematics, and in which—in contrast to the purely formal modes of inference in mathematics proper—one applies contextual inference, but only to prove the consistency of the axioms . . . Thus the development

> of mathematical science as a whole takes place in two ways that constantly alternate: on the one hand we derive provable formulae from the axioms by formal inference; on the other, we adjoin new axioms and prove their consistency by contextual inference.

Hilbert's program sacrificed meaning to secure the foundations. Consequently, to his formalist followers, mathematics was indeed just a game, but their aim was to rigorously prove it to be a fully consistent game. With all the developments in axiomatization, the realization of this formalist "proof-theoretic" dream appeared to be just around the corner.

Not all were convinced, however, that the path taken by Hilbert was the right one. Ludwig Wittgenstein (1889–1951), considered by some to be the greatest philosopher of the twentieth century, regarded Hilbert's efforts with metamathematics as a waste of time. "We cannot lay down a rule for the application of another rule," he argued. In other words, Wittgenstein did not believe that the understanding of one "game" could depend on the construction of another: "If I am unclear about the nature of mathematics, no proof can help me." Still, no one was expecting the lightning that was about to strike. With one blow, the twenty-four-year-old Kurt Gödel would drive a stake right through the heart of formalism.

Kurt Gödel (Figure 53) was born on April 28, 1906, in the Moravian city later known by the Czech name of Brno. At the time, the city was part of the Austro-Hungarian Empire, and Gödel grew up in a German-speaking family. His father, Rudolf Gödel, managed a textile factory and his mother, Marianne Gödel, took care that the young Kurt got a broad education in mathematics, history, languages, and religion. During his teen years, Gödel developed an interest in mathematics and philosophy, and at age eighteen he entered the University of Vienna, where his attention turned primarily to mathematical logic. He was particularly fascinated by Russell and Whitehead's *Principia Mathematica* and by Hilbert's program, and chose for the topic of his dissertation the problem of *completeness*. The goal of this investigation was basically to determine whether the formal approach

Figure 53

advocated by Hilbert was sufficient to produce all the true statements of mathematics. Gödel was awarded his doctorate in 1930, and just one year later he published his *incompleteness theorems,* which sent shock waves through the mathematical and philosophical worlds.

In pure mathematical language, the two theorems sounded rather technical, and not particularly exciting:

1. Any consistent formal system *S* within which a certain amount of elementary arithmetic can be carried out is incomplete with regard to statements of elementary arithmetic: there are such statements which can neither be proved nor disproved in *S*.
2. For any consistent formal system *S* within which a certain amount of elementary arithmetic can be carried out, the consistency of *S* cannot be proved in *S* itself.

The words may appear to be benign, but the implications for the formalists' program were far-reaching. Put somewhat simplistically, the incompleteness theorems proved that Hilbert's formalist program was essentially doomed from the start. Gödel showed that any formal system that is powerful enough to be of any interest is inherently either incomplete or inconsistent. That is, in the best case, there will always be assertions that the formal system can neither prove nor disprove. In the worst, the system would yield contradictions. Since it is always the case that for any statement *T*, either *T* or not-*T* has to be true, the fact that a finite formal system can neither prove nor disprove certain assertions means that true statements will always exist that are not provable within the system. In other words, Gödel demonstrated that no formal system composed of a finite set of axioms and rules of inference can *ever* capture the entire body of truths of mathematics. The most one can hope for is that the commonly accepted axiomatizations are only incomplete and not inconsistent.

Gödel himself believed that an independent, Platonic notion of mathematical truth did exist. In an article published in 1947 he wrote:

> But, despite their remoteness from sense experience, we do have something like a perception of the objects of set theory, as is seen from the fact that the axioms force themselves upon us as being true. I don't see any reason why we should have less confidence in this kind of perception, i.e., in mathematical intuition, than in sense perception.

By an ironic twist of fate, just as the formalists were getting ready for their victory march, Kurt Gödel—an avowed Platonist—came and rained on the parade of the formalist program.

The famous mathematician John von Neumann (1903–57), who was lecturing on Hilbert's work at the time, canceled the rest of his planned course and devoted the remaining time to Gödel's findings.

Gödel the man was every bit as complex as his theorems. In 1940, he and his wife Adele fled Nazi Austria so he could take up a position at the Institute for Advanced Study in Princeton, New Jersey. There he became a good friend and walking partner of Albert Einstein. When Gödel applied for naturalization as an American citizen in 1948, it was Einstein who, together with Princeton University mathematician and economist Oskar Morgenstern (1902–77), accompanied Gödel to his interview at the Immigration and Naturalization Service office. The events surrounding this interview are generally known, but they are so revealing about Gödel's personality that I will give them now in full, *precisely* as they were recorded from memory by Oskar Morgenstern on September 13, 1971. I am grateful to Ms. Dorothy Morgenstern Thomas, Morgenstern's widow, and to the Institute for Advanced Study for providing me with a copy of the document:

> It was in 1946 that Gödel was to become an American citizen. He asked me to be his witness and as the other witness, he proposed Albert Einstein who also gladly consented. Einstein and I occasionally met and were full of anticipation as to what would happen during this time prior to the naturalization proceedings themselves and even during those.
>
> Gödel whom I have seen of course time and again in the months before this event began to go in a thorough manner to prepare himself properly. Since he is a very thorough man, he started informing himself about the history of the settlement of North America by human beings. That led gradually to the study of the History of American Indians, their various tribes, etc. He called me many times on the phone to get literature which he diligently perused. There were many questions raised gradually and of course many doubts brought forth as to whether these histories really were correct and what peculiar circumstances were revealed in them. From that, Gödel gradually over the next weeks proceeded to study American history, concentrating in particular on matters of constitutional law. That also led him into the study of Princeton, and he wanted to know from me in particular where the borderline was between the borough and the township. I tried to explain that all this was totally unnecessary, of course, but with no avail. He persisted in finding out all the facts he wanted to know about and so I provided him with the proper information, also about Princeton. Then he wanted to know how the Borough Council was elected, the Township Council, and who the Mayor was, and how the Township Council functioned. He thought he might be asked about such matters. If he were to show that

he did not know the town in which he lived, it would make a bad impression.

I tried to convince him that such questions never were asked, that most questions were truly formal and that he would easily answer them; that at most they might ask what sort of government we have in this country or what the highest court is called, and questions of this kind. At any rate, he continued with the study of the Constitution.

Now came an interesting development. He rather excitedly told me that in looking at the Constitution, to his distress, he had found some inner contradictions and that he could show how in a perfectly legal manner it would be possible for somebody to become a dictator and set up a Fascist regime, never intended by those who drew up the Constitution. I told him that it was most unlikely that such events would ever occur, even assuming that he was right, which of course I doubted. But he was persistent and so we had many talks about this particular point. I tried to persuade him that he should avoid bringing up such matters at the examination before the court in Trenton, and I also told Einstein about it: he was horrified that such an idea had occurred to Gödel, and he also told him he should not worry about these things nor discuss that matter.

Many months went by and finally the date for the examination in Trenton came. On that particular day, I picked up Gödel in my car. He sat in the back and then we went to pick up Einstein at his house on Mercer Street, and from there we drove to Trenton. While we were driving, Einstein turned around a little and said, "Now, Gödel, are you *really* well prepared for this examination?" Of course, this remark upset Gödel tremendously, which was exactly what Einstein intended and he was greatly amused when he saw the worry on Gödel's face. When we came to Trenton, we were ushered into a big room, and while normally the witnesses are questioned separately from the candidate, because of Einstein's appearance, an exception was made and all three of us were invited to sit down together, Gödel, in the center. The examinor [*sic*] first asked Einstein and then me whether we thought Gödel would make a good citizen. We assured him that this would certainly be the case, that he was a distinguished man, etc. And then he turned to Gödel and said, "Now, Mr. Gödel, where do you come from?"

Gödel: Where I come from? Austria.

The Examinor: What kind of government did you have in Austria?

Gödel: It was a republic, but the constitution was such that it finally was changed into a dictatorship.

The Examinor: Oh! This is very bad. This could not happen in this country.

Gödel: Oh, yes, I can *prove* it.

So of all the possible questions, just that critical one was asked by the Examinor. Einstein and I were horrified during this exchange; the Examinor was intelligent enough to quickly quieten Gödel and say, "Oh God, let's not go into this," and broke off the examination at this point, greatly to our relief. We finally left, and as we were walking out towards the elevators, a man came running after us with a piece of paper and a pen and approached Einstein and asked him for his autograph. Einstein obliged. When we went down in the elevator, I turned to Einstein and said, "It must be dreadful to be persecuted in this fashion by so many people." Einstein said to me, "You know, this is just the last remnant of cannibalism." I was puzzled and said, "How is that?" He said: "Yes, formerly they wanted your blood, now they want your ink."

Then we left, drove back to Princeton, and as we came to the corner of Mercer Street, I asked Einstein whether he wanted to go to the Institute or home. He said, "Take me home, my work is not worth anything anyway anymore." Then he quoted from an American political song (I unfortunately do not recall the

> words, I may have it in my notes and I would certainly recognize it if somebody would suggest the particular phrase). Then off to Einstein's home again, and then he turned back once more toward Gödel, and said, "Now, Gödel, this was your one but last examination." Gödel: "Goodness, is there still another one to come?" and he was already worried. And then Einstein said, "Gödel, the next examination is when you step into your grave." Gödel: "But Einstein, I don't step into my grave," and then Einstein said, "Gödel, that's just the joke of it!" and with that he departed. I drove Gödel home. Everybody was relieved that this formidable affair was over; Gödel had his head free again to go about problems of philosophy and logic.

Late in life Gödel suffered from periods of serious mental disorder, which resulted in his refusal to eat. He died on January 14, 1978, of malnutrition and exhaustion.

Contrary to some popular misconceptions, Gödel's incompleteness theorems do not imply that some truths will never become known. We also cannot infer from the theorems that the human capacity for understanding is somehow limited. Rather, the theorems only demonstrate the weaknesses and shortcomings of formal systems. It may therefore come as a surprise that in spite of the theorems' broad import for the philosophy of mathematics, their impact on the effectiveness of mathematics as a theory-building machinery has been rather minimal. In fact, during the decades surrounding the publication of Gödel's proof, mathematics was reaching some of its most spectacular successes in physical theories of the universe. Far from being abandoned as unreliable, mathematics and its logical conclusions were becoming increasingly essential for the understanding of the cosmos.

What this meant, however, was that the puzzle of the "unreasonable effectiveness" of mathematics became even thornier. Think about this for a moment. Imagine what would have happened had the logicist endeavor been entirely successful. This would have implied that mathematics stems fully from logic—literally from the laws of thought. But how could such a deductive science so marvelously fit natural phenomena? What is the relation between formal logic (maybe we should even say human formal logic) and the cosmos? The answer did not become any clearer after Hilbert and Gödel. Now all that existed was an incomplete formal "game," expressed in mathematical language. How could models based on such an "unreliable" system produce deep insights about the universe and its workings? Before I even attempt to address these questions, I want to sharpen them a bit further by examining a few case studies that demonstrate the subtleties of the effectiveness of mathematics.

CHAPTER 8

UNREASONABLE EFFECTIVENESS?

In chapter 1, I noted that the success of mathematics in physical theories has two aspects: one I called "active" and one "passive." The "active" side reflects the fact that scientists formulate the laws of nature in manifestly applicable mathematical terms. That is, they use mathematical entities, relations, and equations that were developed with an application in mind, often for the very topic under discussion. In those cases the researchers tend to rely on the perceived similarity between the properties of the mathematical concepts and the observed phenomena or experimental results. The effectiveness of mathematics may not appear to be so surprising in these cases, since one could argue that the theories were tailored to fit the observations. There is still, however, an astonishing part of the "active" use related to accuracy, which I will discuss later in this chapter. The "passive" effectiveness refers to cases in which entirely abstract mathematical theories had been developed, with no intended application, only to metamorphose later into powerfully predictive physical models. *Knot theory* provides a spectacular example of the interplay between active and passive effectiveness.

Knots

Knots are the stuff that even legends are made of. You may recall the Greek legend of the Gordian knot. An oracle decreed to the citizens of Phrygia that their next king would be the first man to enter the capital in an oxcart. Gordius, an unsuspecting peasant who happened to ride an oxcart into town, thus became king. Overwhelmed with gratitude, Gordius dedicated his wagon to the gods, and he tied it to a pole with an intricate knot that defied all attempts to untie it. A later prophecy pronounced that the person to untie the knot would become king of Asia. As fate would have it, the man who eventually untied the knot (in the year 333 BC) was Alexander the Great, and he indeed subsequently became ruler of Asia. Alexander's solution to the Gordian knot, however, was not exactly one we would call subtle or even fair—he apparently sliced through the knot with his sword!

But we don't have to go all the way back to ancient Greece to encounter knots. A child tying his shoelaces, a girl braiding her hair, a grandma knitting a sweater, or a sailor mooring a boat are all using knots of some sort. Various knots were even given imaginative names, such as "fisherman's bend," "Englishman's tie," "cat's paw," "true-lover's knot," "granny," and "hangman's knot." Maritime knots in particular were historically considered sufficiently important to have inspired an entire collection of books about them in seventeenth century England. One of those books, incidentally, was written by none other than the English adventurer John Smith (1580–1631), better known for his romantic relationship with the native American princess Pocahontas.

The mathematical theory of knots was born in 1771 in a paper written by the French mathematician Alexandre-Théophile Vandermonde (1735–96). Vandermonde was the first to recognize that knots could be studied as part of the subject of *geometry of position,* which deals with relations depending on position alone, ignoring sizes and calculation with quantities. Next in line, in terms of his role in the development of knot theory, was the German "Prince of Mathematics," Carl Friedrich Gauss. Several of Gauss's notes contain drawings and detailed descriptions of knots, along with some analytic examinations of their properties. As important as the works of Vandermonde, Gauss, and a few other nineteenth century mathematicians were, however, the main driving force behind the modern mathematical knot theory came from an unexpected source—an attempt to explain the structure of matter. The idea originated in the mind of the famous

English physicist William Thomson, better known today as Lord Kelvin (1824–1907). Thomson's efforts concentrated on formulating a theory of atoms, the basic building blocks of matter. According to his truly imaginative conjecture, atoms were really knotted tubes of ether—that mysterious substance that was supposed to permeate all space. The variety of chemical elements could, in the context of this model, be accounted for by the rich diversity of knots.

If Thomson's speculation sounds almost crazy today, it is only because we have had an entire century to get used to and test experimentally the correct model of the atom, in which electrons orbit the atomic nucleus. But this was England of the 1860s, and Thomson was deeply impressed with the stability of complex smoke rings and their ability to vibrate—two properties considered essential for modeling atoms at the time. In order to develop the knot equivalent of a periodic table of the elements, Thomson had to be able to classify knots—find out which different knots are possible—and it was this need for knot tabulation that sparked a serious interest in the mathematics of knots.

As I explained already in chapter 1, a mathematical knot looks like a familiar knot in a string, only with the string's ends spliced. In other words, a mathematical knot is portrayed by a closed curve with no loose ends. A few examples are presented in figure 54, where the three-dimensional knots are represented by their projections, or shadows, in the plane. The position in space of any two strands that cross each other is indicated in the figure by interrupting the line that depicts the lower strand. The simplest knot—the one called the *unknot*—is just a closed circular curve (as in figure 54a). The *trefoil knot* (shown in figure 54b) has three crossings of the strands, and the *figure eight knot* (figure 54c) has four crossings. In Thomson's theory, these three knots could, in principle, be models of three atoms of increasing complexity, such as the hydrogen, carbon, and oxygen atoms, respectively. Still, a complete knot classification was badly needed, and the person who set out to sort the knots was Thomson's friend the Scottish mathematical physicist Peter Guthrie Tait (1831–1901).

The types of questions mathematicians ask about knots are really not very different from those one might ask about an ordinary knotted string or a tangled ball of yarn. Is it really knotted? Is one knot

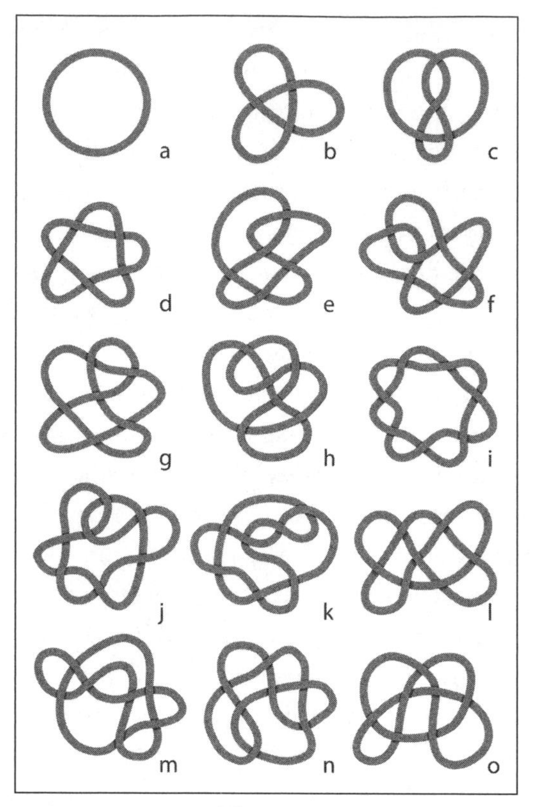

Figure 54

equivalent to another? What the latter question means is simply: Can one knot be deformed into the shape of the other without breaking the strands or pushing one strand through the other like a magician's linking rings? The importance of this question is demonstrated in figure 55, which shows that by certain manipulations one can obtain two very different representations of what is actually the same knot. Ultimately, knot theory searches for some precise way of proving that certain knots (such as the trefoil knot and the figure eight knot; figures 54b and 54c) are really different, while ignoring the superficial differences of other knots, such as the two knots in figure 55.

Tait started his classification work the hard way. Without any rigorous mathematical principle to guide him, he compiled lists of curves with one crossing, two crossings, three crossings, and so on. In collaboration with the Reverend Thomas Penyngton Kirkman

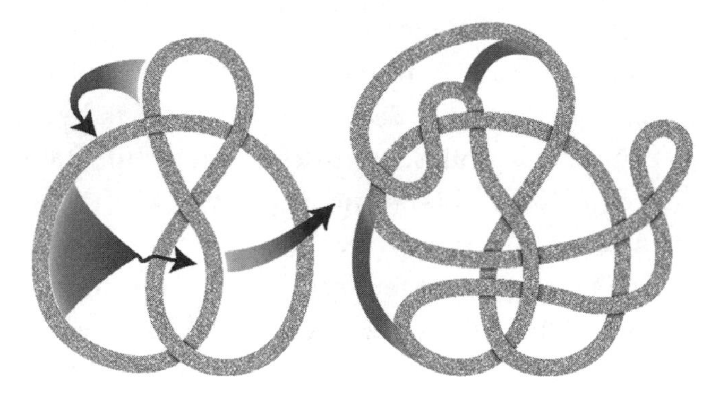

Figure 55

(1806–95), who was also an amateur mathematician, he started sifting through the curves to eliminate duplications by equivalent knots. This was not a trivial task. You must realize that at every crossing, there are two ways to choose which strand would be uppermost. This means that if a curve contains, say, seven crossings, there are 2 × 2 × 2 × 2 × 2 × 2 × 2 = 128 knots to consider. In other words, human life is too short to complete in this intuitive way the classification of knots with tens of crossings or more. Nevertheless, Tait's labor did not go unappreciated. The great James Clerk Maxwell, who formulated the classical theory of electricity and magnetism, treated Thomson's atomic theory with respect, stating that "it satisfies more of the conditions than any atom hitherto considered." Being at the same time well aware of Tait's contribution, Maxwell offered the following rhyme:

Clear your coil of kinkings
Into perfect plaiting,
Locking loops and linkings
Interpenetrating.

By 1877, Tait had classified alternating knots with up to seven crossings. Alternating knots are those in which the crossings go alternately over and under, like the thread in a woven carpet. Tait also made a few more pragmatic discoveries, in the form of basic principles that were later christened *Tait's conjectures.* These conjectures were so substantial, by the way, that they resisted all attempts to prove them rigorously until the late 1980s. In 1885, Tait published tables of knots with up to ten crossings, and he decided to stop there. Independently, University of Nebraska professor Charles Newton Little (1858–1923) also published (in 1899) tables of nonalternating knots with ten or fewer crossings.

Lord Kelvin always thought fondly of Tait. At a ceremony at Peterhouse College in Cambridge, where a portrait of Tait was presented, Lord Kelvin said:

> I remember Tait once remarking that nothing but science is worth living for. It was sincerely said, but Tait himself proved it to be not true. Tait was a great reader. He would get Shakespeare, Dickens, and Thackeray off by heart. His memory was wonderful. What he once read sympathetically he ever after remembered.

Unfortunately, by the time Tait and Little completed their heroic work on knot tabulation, Kelvin's theory had already been totally discarded as a potential atomic theory. Still, interest in knots continued for its own sake, the difference being that, as the mathematician Michael Atiyah has put it, "the study of knots became an esoteric branch of pure mathematics."

The general area of mathematics where qualities such as size, smoothness, and in some sense even shape are ignored is called *topology*. Topology—the rubber-sheet geometry—examines those properties that remain unchanged when space is stretched or deformed in any fashion (without tearing off pieces or poking holes). By their very nature, knots belong in topology. Incidentally, mathematicians distinguish between *knots,* which are single knotted loops, *links,* which are sets of knotted loops all tangled together, and *braids,* which are sets of vertical strings attached to a horizontal bar at the top and bottom ends.

If you were not impressed with the difficulty of classifying knots, consider the following very telling fact. Charles Little's table, published in 1899 after six years of work, contained forty-three nonalternating knots of ten crossings. This table was scrutinized by many mathematicians and believed to be correct for seventy-five years. Then in 1974, the New York lawyer and mathematician Kenneth Perko was experimenting with ropes on his living room floor. To his surprise, he discovered that two of the knots in Little's table were in fact the same. We now believe that there are only forty-two distinct nonalternating knots of ten crossings.

While the twentieth century witnessed great strides in topology, progress in knot theory was relatively slow. One of the key goals of the mathematicians studying knots has been to identify properties that truly distinguish knots. Such properties are called *invariants of knots*—they represent quantities for which any two different projections of the same knot yield precisely the same value. In other words, an ideal invariant is literally a "fingerprint" of the knot—a characteristic property of the knot that does not change by deformations of the knot. Perhaps the simplest invariant one can think of is the minimum number of crossings in a drawing of the knot. For instance, no matter how hard you try to disentangle the trefoil knot (figure 54b), you will never reduce the number of crossings to fewer than three. Unfortunately, there are a number of reasons why the minimal number of crossings is not the most useful invariant. First, as figure 55 demonstrates, it is not always easy to determine whether a knot has been drawn with the minimum number of crossings. Second and more important, many knots that are actually different have the same number of crossings. In figure 54, for instance, there are three different knots with six crossings, and no fewer than seven different knots with seven crossings. The minimum number of crossings, therefore, does not distinguish most knots. Finally, the minimum number of crossings, by its very simplistic nature, does not provide much insight into the properties of knots in general.

A breakthrough in knot theory came in 1928 when the American mathematician James Waddell Alexander (1888–1971) discovered an important invariant that has become known as the *Alexander polynomial.* Basically, the Alexander polynomial is an algebraic expression that uses the arrangement of crossings to label the knot. The good news was that if two knots had different Alexander polynomials, then the knots were definitely different. The bad news was that two knots that had the same polynomial could still be different knots. While extremely helpful, therefore, the Alexander polynomial was still not perfect for distinguishing knots.

Mathematicians spent the next four decades exploring the conceptual basis for the Alexander polynomial and gaining further insights into the properties of knots. Why were they getting so deeply into that subject? Certainly not for any practical application. Thomson's atomic model had long been forgotten, and there was no other problem in sight in the sciences, economics, architecture, or any other discipline that appeared to require a theory of knots. Mathematicians were spending endless hours on knots simply because they were curious! To these individuals, the idea of understanding knots and the principles that govern them was exquisitely beautiful. The sudden flash of insight afforded by the Alexander polynomial was as irresistible to mathematicians as the challenge of climbing Mount Everest was to George Mallory, who famously replied "Because it is there" to the question of why he wanted to climb the mountain.

In the late 1960s, the prolific English-American mathematician John Horton Conway discovered a procedure for "unknotting" knots gradually, thereby revealing the underlying relationship between knots and their Alexander polynomials. In particular, Conway introduced two simple "surgical" operations that could serve as the basis for defining a knot invariant. Conway's operations, dubbed *flip* and *smoothing,* are described schematically in figure 56. In the flip (figure 56a), the crossing is transformed by running the upper strand under the lower one (the figure also indicates how one might achieve this transformation in a real knot in a string). Note that the flip obviously changes the nature of the knot. For instance, you can easily convince yourself that the trefoil knot in figure 54b would become the unknot (figure 54a) following a flip. Conway's smoothing operation eliminates the crossing altogether (figure 56b), by reattaching the strands the "wrong" way. Even with the new understanding gained from Conway's work, mathematicians remained convinced for almost two more decades that no other knot invariants (of the type of the Alexander polynomial) could be found. This situation changed dramatically in 1984.

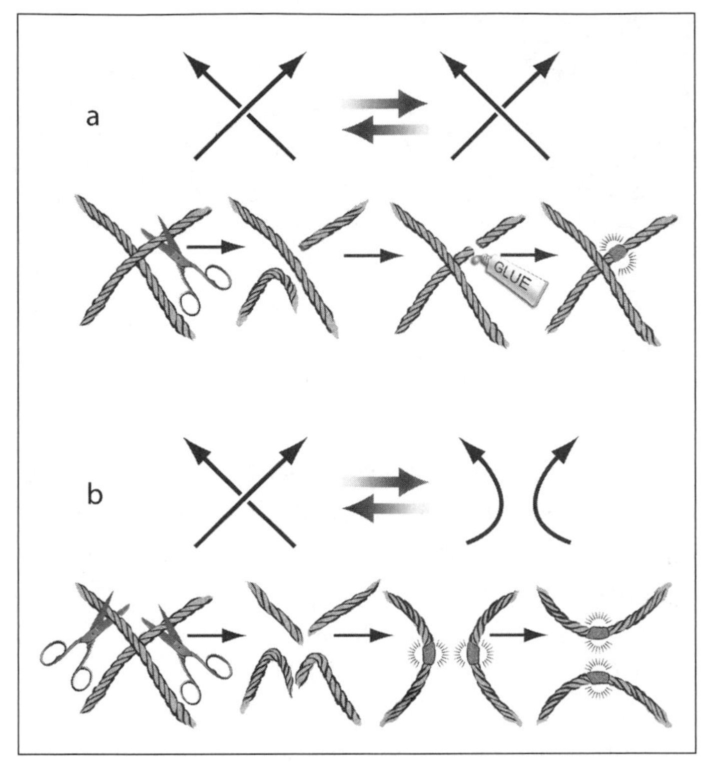

Figure 56

The New Zealander–American mathematician Vaughan Jones was not studying knots at all. Rather, he was exploring an even more abstract world—one of the mathematical entities known as *von Neumann algebras*. Unexpectedly, Jones noticed that a relation that surfaced in von Neumann algebras looked suspiciously similar to a relation in knot theory, and he met with Columbia University knot theorist Joan Birman to discuss possible applications. An examination of that relation eventually revealed an entirely new invariant for knots, dubbed the *Jones polynomial*. The Jones polynomial was immediately recognized as a more sensitive invariant than the Alexander polynomial. It distinguishes, for instance, between knots and their mirror images (e.g., the right-handed and left-handed trefoil knots in figure 57), for which the Alexander polynomials were identical. More importantly, however, Jones's discovery generated an unprecedented excitement among knot theorists. The announcement

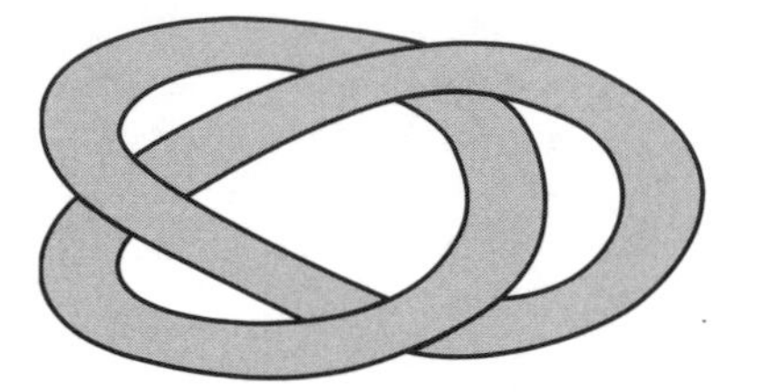

Figure 57

of a new invariant triggered such a flurry of activity that the world of knots suddenly resembled the stock exchange floor on a day on which the Federal Reserve unexpectedly lowers interest rates.

There was much more to Jones's discovery than just progress in knot theory. The Jones polynomial suddenly connected a bewildering variety of areas in mathematics and physics, ranging from statistical mechanics (used, for instance, to study the behavior of large collections of atoms or molecules) to quantum groups (a branch of mathematics related to the physics of the subatomic world). Mathematicians all over the world immersed themselves feverishly in attempts to look for even more general invariants that would somehow encompass both the Alexander and Jones polynomials. This mathematical race ended up in what is perhaps the most astonishing result in the history of scientific competition. Only a few months after Jones revealed his new polynomial, four groups, working independently and using three different mathematical approaches, announced *at the same time* the discovery of an even more sensitive invariant. The new polynomial became known as the *HOMFLY polynomial*, after the first letters in the names of the discoverers: Hoste, Ocneanu, Millett, Freyd, Lickorish, and Yetter. Furthermore, as if four groups crossing the finish line in a dead heat weren't enough, two Polish mathematicians (Przytycki and Traczyk) discovered independently precisely the same polynomial, but they missed the publication date due to a capricious mail system. Consequently, the polynomial is also referred to as the HOMFLYPT (or sometimes THOMFLYP) polynomial, adding the first letters in the names of the Polish discoverers.

Since then, while other knot invariants have been discovered, a complete classification of knots remains elusive. The question of precisely which knot can be twisted and turned to produce another knot without the use of scissors is still unanswered. The most advanced invariant discovered to date is the work of the Russian-French mathematician Maxim Kontsevich, who received the prestigious Fields Medal in 1998 and the Crafoord Prize in 2008 for his work. Incidentally, in 1998, Jim Hoste of Pitzer College in Claremont, California, and Jeffrey Weeks of Canton, New York, tabulated all the knotted loops having sixteen or fewer crossings. An identical tabulation was produced independently by Morwen Thistlethwaite of the University of Tennessee in Knoxville. Each list contains precisely 1,701,936 different knots!

The real surprise, however, came not so much from the progress in knot theory itself, but from the dramatic and unexpected comeback that knot theory has made in a wide range of sciences.

The Knots of Life

Recall that knot theory was motivated by a wrong model of the atom. Once that model died, however, mathematicians were not discouraged. On the contrary, they embarked with great enthusiasm on the long and difficult journey of trying to understand knots in their own right. Imagine then their delight when knot theory suddenly turned out to be the key to understanding fundamental processes involving the molecules of life. Do you need any better example of the "passive" role of pure mathematics in explaining nature?

Deoxyribonucleic acid, or DNA, is the genetic material of all cells. It consists of two very long strands that are intertwined and twisted around each other millions of times to form a double helix. Along the two backbones, which can be thought of as the sides of a ladder, sugar and phosphate molecules alternate. The "rungs" of the ladder consist of pairs of bases connected by hydrogen bonds in a prescribed fashion (adenine bonds only with thymine, and cytosine only with guanine; figure 58). When a cell divides, the first step is replication of DNA, so that daughter cells can receive copies. Similarly, in the process of *transcription* (in which genetic information from DNA is copied to RNA), a section of the DNA double helix is uncoiled and only one DNA

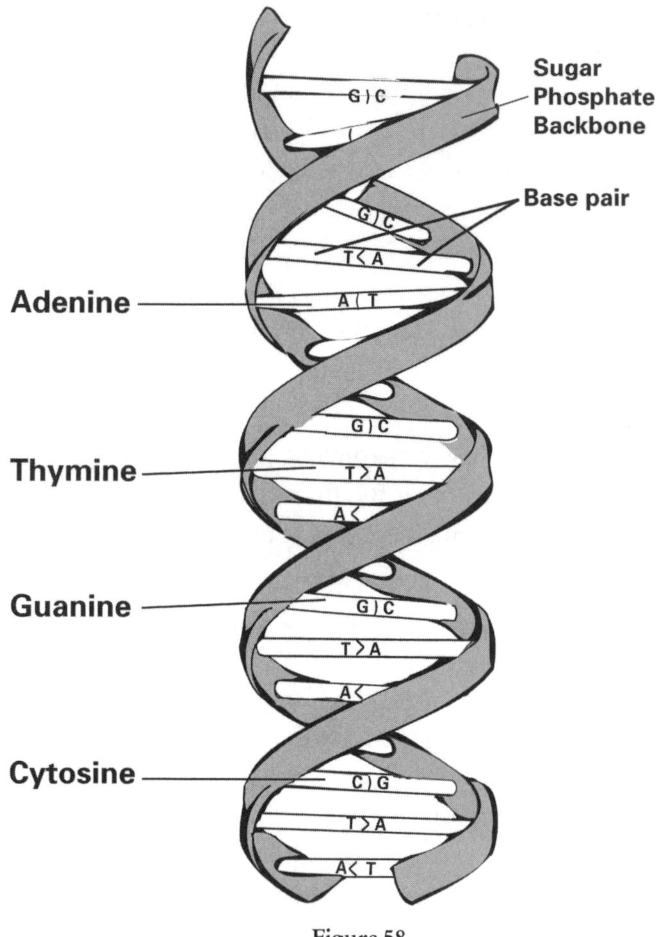

Figure 58

strand serves as a template. After the synthesis of RNA is complete, the DNA recoils into its helix. Neither the replication nor the transcription process is easy, however, because DNA is so tightly knotted and coiled (in order to compact the information storage) that unless some unpacking takes place, these vital life processes could not proceed smoothly. In addition, for the replication process to reach completion, offspring DNA molecules must be unknotted, and the parent DNA must eventually be restored to its original configuration.

The agents that take care of the unknotting and disentanglement are enzymes. Enzymes can pass one DNA strand through another by creating temporary breaks and reconnecting the ends differently. Does this process sound familiar? These are precisely the surgical operations introduced by Conway for the unraveling of mathematical knots (represented in figure 56). In other words, from a topological standpoint, DNA is a complex knot that has to be unknotted by enzymes to allow for replication or transcription to occur. By using knot theory to calculate how difficult it is to unknot the DNA, researchers can study the properties of the enzymes that do the unknotting. Better yet, using experimental visualization techniques such as electron microscopy and gel electrophoresis, scientists can actually observe and quantify the changes in the knotting and linking of DNA caused by an enzyme (figure 59 shows an electron micrograph of a DNA knot). The challenge to mathematicians is then to deduce the mechanisms by which the enzymes operate from the observed changes in the topology of the DNA. As a byproduct, the changes in the number of crossings in the DNA knot give biologists a measure of the *reaction rates* of the enzymes—how many crossings per minute can an enzyme of a given concentration affect.

But molecular biology is not the only arena in which knot theory

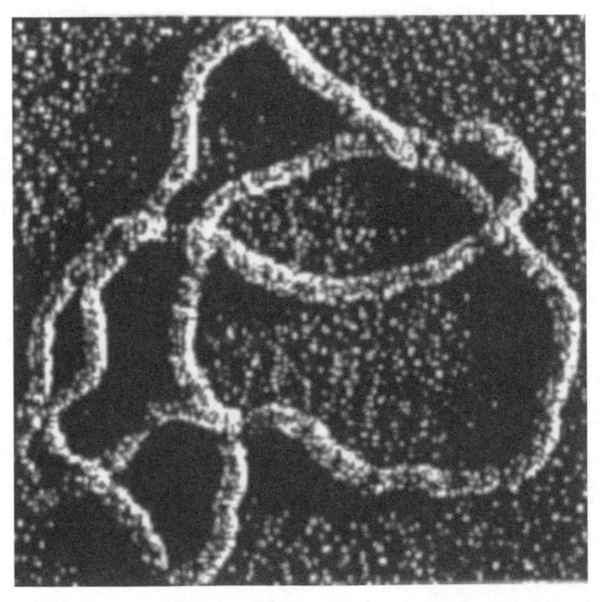

Figure 59

found unforeseen applications. String theory—the current attempt to formulate a unified theory that explains all the forces in nature—is also concerned with knots.

The Universe on a String?

Gravity is the force that operates on the largest scales. It holds the stars in the galaxies together, and it influences the expansion of the universe. Einstein's general relativity is a remarkable theory of gravity. Deep within the atomic nucleus, other forces and a different theory reign supreme. The strong nuclear force holds particles called *quarks* together to form the familiar protons and neutrons, the basic constituents of matter. The behavior of the particles and the forces in the subatomic world is governed by the laws of quantum mechanics. Do quarks and galaxies play by the same rules? Physicists believe they should, even though they don't yet quite know why. For decades, physicists have been searching for a "theory of everything"—a comprehensive description of the laws of nature. In particular, they want to bridge the gap between the large and the small with a quantum theory of gravity—a reconciliation of general relativity with quantum mechanics. String theory appears to be the current best bet for such a theory of everything. Originally developed and discarded as a theory for the nuclear force itself, string theory was revived from obscurity in 1974 by physicists John Schwarz and Joel Scherk. The basic idea of string theory is quite simple. The theory proposes that elementary subatomic particles, such as electrons and quarks, are not pointlike entities with no structure. Rather, the elementary particles represent different modes of vibration of the same basic string. The cosmos, according to these ideas, is filled with tiny, flexible, rubber band–like loops. Just as a violin string can be plucked to produce different harmonies, different vibrations of these looping strings correspond to distinct matter particles. In other words, the world is something like a symphony.

Since strings are closed loops moving through space, as time progresses, they sweep areas (known as *world sheets*) in the form of cylinders (as in figure 60). If a string emits other strings, this cylinder

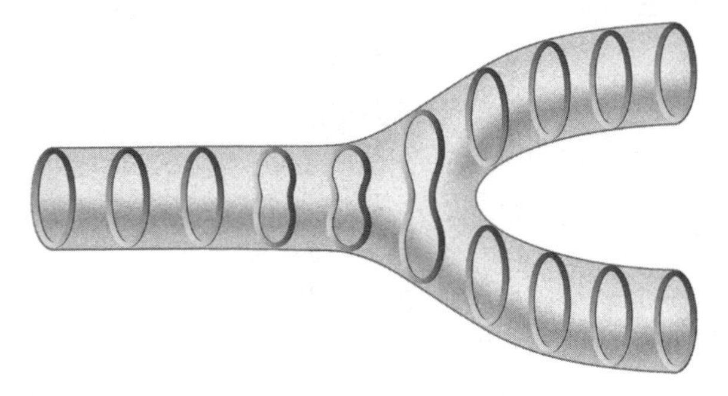

Figure 60

forks to form wishbone-shaped structures. When many strings interact, they form an intricate network of fused donutlike shells. While studying these types of complex topological structures, string theorists Hirosi Ooguri and Cumrun Vafa discovered a surprising connection between the number of donut shells, the intrinsic geometric properties of knots, and the Jones polynomial. Even earlier, Ed Witten—one of the key players in string theory—created an unexpected relation between the Jones polynomial and the very foundation of string theory (known as *quantum field theory*). Witten's model was later rethought from a purely mathematical perspective by the mathematician Michael Atiyah. So string theory and knot theory live in perfect symbiosis. On one hand, string theory has benefited from results in knot theory; on the other, string theory has actually led to new insights in knot theory.

With a much broader scope, string theory searches for explanations for the most basic constituents of matter, much in the same way that Thomson originally searched for a theory of atoms. Thomson (mistakenly) thought that knots could provide the answer. By a surprising twist, string theorists find that knots can indeed provide at least some answers.

The story of knot theory demonstrates beautifully the unexpected powers of mathematics. As I have mentioned earlier, even the "active" side of the effectiveness of mathematics alone—when scientists generate the mathematics they need to describe observable science—presents some baffling surprises when it comes to accuracy. Let me describe briefly one topic in physics in which both the active and the passive aspects played a role, but which is particularly remarkable because of the obtained accuracy.

A Weighty Accuracy

Newton took the laws of falling bodies discovered by Galileo and other Italian experimentalists, combined them with the laws of planetary motion determined by Kepler, and used this unified scheme to put forth a universal, mathematical law of gravitation. Along the way, Newton had to formulate an entirely new branch of mathematics—calculus—that allowed him to capture concisely and coherently all the properties of his proposed laws of motion and gravitation. The accuracy to which Newton himself could verify his law of gravity, given the experimental and observational results of his day, was no better than about 4 percent. Yet the law proved to be accurate beyond all reasonable expectations. By the 1950s the experimental accuracy was better than one ten-thousandth of a percent. But this is not all. A few recent, speculative theories, aimed at explaining the fact that the expansion of our universe seems to be speeding up, suggested that gravity may change its behavior on very small distance scales. Recall that Newton's law states that the gravitational attraction decreases as the inverse square of the distance. That is, if you double the distance between two masses, the gravitational force each mass feels becomes four times weaker. The new scenarios predicted deviations from this behavior at distances smaller than one millimeter (the twenty-fifth part of an inch). Eric Adelberger, Daniel Kapner, and their collaborators at the University of Washington, Seattle, conducted a series of ingenious experiments to test this predicted change in the dependence on the separation. Their most recent results, published in January 2007, show that the inverse-square law holds down to a distance of fifty-six thousandths of a millimeter! So a mathematical law that was proposed more than three hundred years ago on the basis of very scanty observations not only turned out to be phenomenally accurate, but also proved to hold in a range that couldn't even be probed until very recently.

There was one major question that Newton left completely unanswered: How does gravity really work? How does the Earth, a quarter million miles away from the Moon, affect the Moon's motion? Newton was aware of this deficiency in his theory, and he openly admitted it in the *Principia:*

> Hitherto we have explained the phenomena of the heavens and of our sea by the power of gravity, but have not yet assigned the cause of this power. This is certain, that it must proceed from a cause that penetrates to the very centres of the Sun and planets . . . and propagates its virtue on all sides to immense distances, decreasing always as the inverse square of the distances . . . But hitherto I have not been able to discover the cause of those properties of gravity from phenomena, and I frame no hypotheses.

The person who decided to meet the challenge posed by Newton's omission was Albert Einstein (1879–1955). In 1907 in particular, Einstein had a very strong reason to be interested in gravity—his new theory of *special relativity* appeared to be in direct conflict with Newton's law of gravitation.

Newton believed that gravity's action was instantaneous. He assumed that it took no time at all for planets to feel the Sun's gravitational force, or for an apple to feel the Earth's attraction. On the other hand, the central pillar of Einstein's special relativity was the statement that no object, energy, or information could travel faster than the speed of light. So how could gravity work instantaneously? As the following example will show, the consequences of this contradiction could be disastrous to concepts as fundamental as our perception of cause and effect.

Imagine that the Sun were to somehow suddenly disappear. Robbed of the force holding it to its orbit, the Earth would (according to Newton) immediately start moving along a straight line (apart from small deviations due to the gravity of the other planets). However, the Sun would actually disappear from view to the Earth's inhabitants only about eight minutes later, since this is the time it takes light to traverse the distance from the Sun to the Earth. In other words, the change in the Earth's motion would precede the Sun's disappearance.

To remove this conflict, and at the same time to tackle Newton's unanswered question, Einstein engaged almost obsessively in a search for a new theory of gravity. This was a formidable task. Any new theory had not only to preserve all the remarkable successes of Newton's theory, but also to explain how gravity works, and to do so in a way that is compatible with special relativity. After a number of false starts and long wanderings down blind alleys, Einstein finally reached his goal in 1915. His *theory of general relativity* is still regarded by many as one of the most beautiful theories ever formulated.

At the heart of Einstein's groundbreaking insight lay the idea that gravity is nothing but warps in the fabric of space and time. According to Einstein, just as golf balls are guided by the warps and curves across an undulating green, planets follow curved paths in the warped space representing the Sun's gravity. In other words, in the absence of matter or other forms of energy, *spacetime* (the unified fabric of the three dimensions of space and one of time) would be flat. Matter and energy warp spacetime just as a heavy bowling ball causes a trampoline to sag. Planets follow the most direct paths in this curved geometry, which is a manifestation of gravity. By solving the "how it works" problem for gravity, Einstein also provided the framework for addressing the question of how fast it propagates. The latter question boiled down to determining how fast warps in spacetime could travel. This was a bit like calculating the speed of ripples in a pond. Einstein was able to show that in general relativity gravity traveled precisely at the speed of light, which eliminated the discrepancy that existed between Newton's theory and special relativity. If the Sun were to disappear, the change in the Earth's orbit would occur eight minutes later, coinciding with our observing the disappearance.

The fact that Einstein had turned warped four-dimensional spacetime into the cornerstone of his new theory of the cosmos meant that he badly needed a mathematical theory of such geometrical entities. In desperation, he turned to his old classmate the mathematician Marcel Grossmann (1878–1936): "I have become imbued with great respect for mathematics, the more subtle parts of which I had previously regarded as sheer luxury." Grossmann pointed out that Riemann's non-Euclidean geometry (described in chapter 6) was precisely the tool that Einstein needed—a geometry of curved spaces in any number of dimensions. This was an incredible demonstration of what I dubbed the "passive" effectiveness of mathematics, which Einstein was quick to acknowledge: "We may in fact regard [geometry] as the most ancient branch of physics," he declared. "Without it I would have been unable to formulate the theory of relativity."

General relativity has also been tested with impressive accuracy. These tests are not easy to come by, since the curvature in spacetime introduced by objects such as the Sun is measured only in parts per million. While the original tests were all associated with observations within the solar system (e.g., tiny changes to the orbit of the planet Mercury, as compared to the predictions of Newtonian gravity), more exotic tests have recently become feasible. One of the best verifications uses an astronomical object known as a *double pulsar.*

A pulsar is an extraordinarily compact, radio-wave-emitting star, with a mass somewhat larger than the mass of the Sun but a radius of only about six miles. The density of such a star (known as a *neutron star*) is so high that one cubic inch of its matter has a mass of about a billion tons. Many of these neutron stars spin very fast, while emitting radio waves from their magnetic poles. When the magnetic axis is somewhat inclined to the rotation axis (as in figure 61), the radio beam from a given pole may cross our line of sight only once every rotation, like the flash of light from a lighthouse. In this case, the radio emission will appear to be pulsed—hence the name "pulsar." In one case, two pulsars revolve around their mutual center of gravity in a close orbit, creating a double-pulsar system.

There are two properties that make this double pulsar an excellent laboratory for testing general relativity: (1) Radio pulsars are superb clocks—their rotation rates are so stable that in fact they surpass atomic clocks in accuracy; and (2) Pulsars are so compact that their gravitational fields are very strong, producing significant relativistic effects. These features allow astronomers to measure very precisely changes in the light travel time from the pulsars to Earth caused by the orbital motion of the two pulsars in each other's gravitational field.

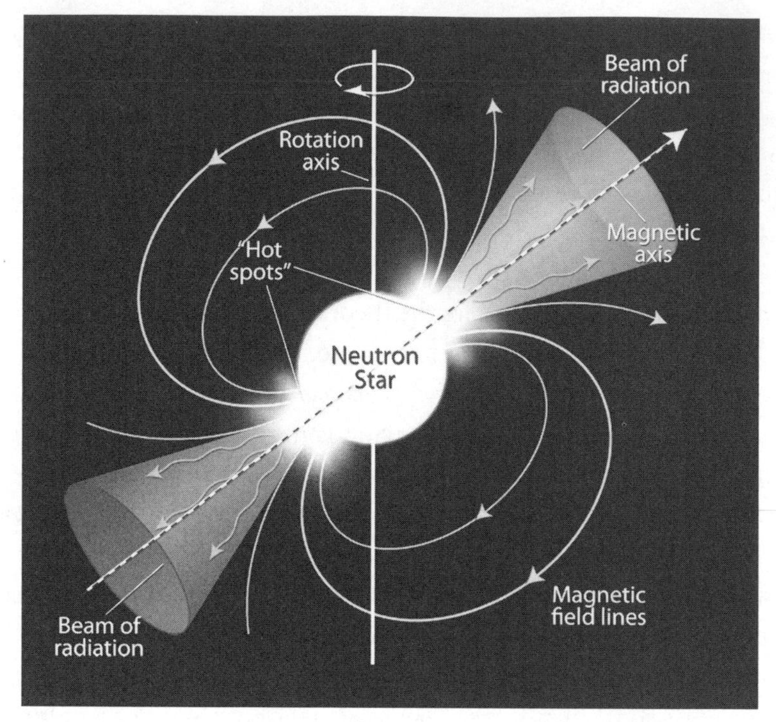

Figure 61

The most recent test was the result of precision timing observations taken over a period of two and a half years on the double-pulsar system known as PSR J0737-3039A/B (the long "telephone number" reflects the coordinates of the system in the sky). The two pulsars in this system complete an orbital revolution in just two hours and twenty-seven minutes, and the system is about two thousand light-years away from Earth (a light-year is the distance light travels in one year in a vacuum; about six trillion miles). A team of astronomers led by Michael Kramer of the University of Manchester measured the relativistic corrections to the Newtonian motion. The results, published in October 2006, agreed with the values predicted by general relativity within an uncertainty of 0.05 percent!

Incidentally, both special relativity and general relativity play an important role in the *Global Positioning System (GPS)* that helps us find our location on the surface of the Earth and our way from place to place, whether in a car, airplane, or on foot. The GPS determines the current position of the receiver by measuring the time it takes the signal from several satellites to reach it and by triangulating on the known positions of each satellite. Special relativity predicts that the atomic clocks on board the satellites should tick more slowly (falling behind by a few millionths of a second per day) than those on the ground because of their relative motion. At the same time, general relativity predicts that the satellite clocks should tick faster (by a few tens of millionths of a second per day) than those on the ground due to the fact that high above the Earth's surface the curvature in spacetime resulting from the Earth's mass is smaller. Without making the necessary corrections for these two effects, errors in global positions could accumulate at a rate of more than five miles in each day.

The theory of gravity is only one of the many examples that illustrate the miraculous suitability and astonishing accuracy of the mathematical formulation of the laws of nature. In this case, as in numerous others, what we got out of the equations was much more than what was originally put in. The accuracy of both Newton's and Einstein's theories proved to far exceed the accuracy of the observations that the theories attempted to explain in the first place.

Perhaps the best example of the astonishing accuracy that a mathematical theory can achieve is provided by *quantum electrodynamics (QED)*, the theory that describes all phenomena involving electrically charged particles and light. In 2006 a group of physicists at Harvard University determined the magnetic moment of the electron (which measures how strongly the electron interacts with a magnetic field) to a precision of eight parts in a trillion. This is an incredible experimental feat in its own right. But when you add to that the fact that the most recent theoretical calculations based on QED reach a similar precision and that the two results agree, the accuracy becomes almost unbelievable. When he heard about the continuing success of QED, one of QED's originators, the physicist Freeman Dyson, reacted: "I'm amazed at how precisely Nature dances to the tune we scribbled so carelessly fifty-seven years ago, and at how the experimenters and the theorists can measure and calculate her dance to a part in a trillion."

But accuracy is not the only claim to fame of mathematical theories—predictive power is another. Let me give just two simple examples, one from the nineteenth century and one from the twentieth century, that demonstrate this potency. The former theory predicted a new phenomenon and the latter the existence of new fundamental particles.

James Clerk Maxwell, who formulated the classical theory of electromagnetism, showed in 1864 that the theory predicted that varying electric or magnetic fields should generate propagating waves. These waves—the familiar electromagnetic waves (e.g., radio)—were first detected by the German physicist Heinrich Hertz (1857–94) in a series of experiments conducted in the late 1880s.

In the late 1960s, physicists Steven Weinberg, Sheldon Glashow, and Abdus Salam developed a theory that treats the electromagnetic force and weak nuclear force in a unified manner. This theory, now known as the *electroweak theory*, predicted the existence of three particles (called the W^+, W^-, and Z bosons) that had never before been observed. The particles were unambiguously detected in 1983 in accelerator experiments (which smash one subatomic particle into another at very high energies) led by physicists Carlo Rubbia and Simon van der Meer.

The physicist Eugene Wigner, who coined the phrase "the unreasonable effectiveness of mathematics," proposed to call all of these unexpected achievements of mathematical theories the "empirical law of epistemology" (epistemology is the discipline that investigates the origin and limits of knowledge). If this "law" were not correct, he argued, scientists would have lacked the encouragement and reassurance that are absolutely necessary for a thorough exploration of the laws of nature. Wigner, however, did not offer any explanation for the empirical law of epistemology. Rather, he regarded it as a "wonderful gift" for which we should be grateful even though we do not understand its origin. Indeed, to Wigner, this "gift" captured the essence of the question about the unreasonable effectiveness of mathematics.

At this point, I believe that we have gathered enough clues that we should at least be able to try answering the questions we started with: Why is mathematics so effective and productive in explaining the world around us that it even yields new knowledge? And, is mathematics ultimately invented or discovered?

CHAPTER 9

ON THE HUMAN MIND, MATHEMATICS, AND THE UNIVERSE

The two questions: (1) Does mathematics have an existence independent of the human mind? and (2) Why do mathematical concepts have applicability far beyond the context in which they have originally been developed? are related in complex ways. Still, to simplify the discussion, I will attempt to address them sequentially.

First, you may wonder where modern-day mathematicians stand on the question of mathematics as a discovery or an invention. Here is how mathematicians Philip Davis and Reuben Hersh described the situation in their wonderful book *The Mathematical Experience:*

> Most writers on the subject seem to agree that the typical working mathematician is a Platonist [views mathematics as discovery] on weekdays and a formalist [views mathematics as invention] on Sundays. That is, when he is doing mathematics he is convinced that he is dealing with an objective reality whose properties he is attempting to determine. But then, when challenged to give a philosophical account of this reality, he finds it easiest to pretend that he does not believe in it after all.

Other than being tempted to substitute "he or she" for "he" everywhere, to reflect the changing mathematical demographics, I have the impression that this characterization continues to be true for many present-day mathematicians and theoretical physicists. Nevertheless, some twentieth century mathematicians did take a strong position on one side or the other. Here, representing the Platonic point of view, is G. H. Hardy in *A Mathematician's Apology:*

> For me, and I suppose for most mathematicians, there is another reality, which I will call "mathematical reality"; and there is no sort of agreement about the nature of mathematical reality among either mathematicians or philosophers. Some hold that it is "mental" and that in some sense we construct it, others that it is outside and independent of us. A man who could give a convincing account of mathematical reality would have solved very many of the most difficult problems of metaphysics. If he could include physical reality in his account, he would have solved them all.
>
> I should not wish to argue any of these questions here even if I were competent to do so, but I will state my own position dogmatically in order to avoid minor misapprehensions. I believe that mathematical reality lies outside us, that our function is to discover or *observe* it, and that the theorems which we prove, and which we describe grandiloquently as our "creations," are simply our notes of our observations. This view has been held, in one form or another, by many philosophers of high reputation from Plato onwards, and I shall use the language which is natural to a man who holds it.

Mathematicians Edward Kasner (1878–1955) and James Newman (1907–66) expressed precisely the opposite perspective in *Mathematics and the Imagination:*

> That mathematics enjoys a prestige unequaled by any other flight of purposive thinking is not surprising. It has made possible so many advances in the sciences, it is at once so indispensable in practical affairs and so easily the masterpiece of pure abstraction that the recognition of its pre-eminence among man's intellectual achievements is no more than its due.

In spite of this pre-eminence, the first significant appraisal of mathematics was occasioned only recently by the advent of non-Euclidean and four-dimensional geometry. That is not to say that the advances made by the calculus, the theory of probability, the arithmetic of the infinite, topology, and the other subjects we have discussed, are to be minimized. Each one has widened mathematics and deepened its meaning as well as our comprehension of the physical universe. Yet none has contributed to mathematical introspection, to the knowledge of the relation of the parts of mathematics to one another and to the whole as much as the non-Euclidean heresies.

As a result of the valiantly critical spirit which engendered the heresies, we have overcome the notion that mathematical truths have an existence independent and apart from our own minds. It is even strange to us that such a notion could ever have existed. Yet this is what Pythagoras would have thought—and Descartes, along with hundreds of other great mathematicians before the nineteenth century. Today mathematics is unbound; it has cast off its chains. Whatever its essence, we recognize it to be as free as the mind, as prehensile as the imagination. Non-Euclidean geometry is proof that mathematics, unlike the music of the spheres, is man's own handiwork, subject only to the limitations imposed by the laws of thought.

So, contrary to the precision and certitude that are the hallmark of statements in mathematics, here we have a divergence of opinions that is more typical of debates in philosophy or politics. Should we be surprised? Not really. The question of whether mathematics is invented or discovered is actually not a question of mathematics at all.

The notion of "discovery" implies preexistence in some universe, either real or metaphysical. The concept of "invention" implicates the human mind, either individually or collectively. The question therefore belongs to a combination of disciplines that may involve physics, philosophy, mathematics, cognitive science, even anthropology, but it is certainly not exclusive to mathematics (at least not directly). Consequently, mathematicians may not even be the best equipped to answer this question. After all, poets, who can perform magic with language, are not necessarily the best linguists, and the greatest philosophers are generally not experts in the functions of the brain. The answer to the "invented or discovered" question can therefore be gleaned only (if at all) from a careful examination of many clues, deriving from a wide variety of domains.

Metaphysics, Physics, and Cognition

Those who believe that mathematics exists in a universe that is independent of humans still fall into two different camps when it comes to identifying the nature of this universe. First, there are the "true" Platonists, for whom mathematics dwells in the abstract, eternal world of mathematical forms. Then there are those who suggest that mathematical structures are in fact a real part of the natural world. Since I have already discussed pure Platonism and some of its philosophical shortcomings quite extensively, let me elaborate a bit on the latter perspective.

The person who presents what may be the most extreme and most speculative version of the "mathematics as a part of the physical world" scenario is an astrophysicist colleague, Max Tegmark of MIT.

Tegmark argues that "our universe is not just described by mathematics—it *is* mathematics" [emphasis added]. His argument starts with the rather uncontroversial assumption that an external physical reality exists that is independent of human beings. He then proceeds to examine what might be the nature of the ultimate theory of such a reality (what physicists refer to as the "theory of everything"). Since this physical world is entirely independent of humans, Tegmark maintains, its description must be free of any human "baggage" (e.g., human language, in particular). In other words, the final theory cannot include any concepts such as "subatomic particles," "vibrating strings," "warped spacetime," or other humanly conceived constructs. From this presumed insight, Tegmark concludes that the only possible description of the cosmos is one that involves only abstract concepts and the relations among them, which he takes to be the working definition of mathematics.

Tegmark's argument for a mathematical reality is certainly intriguing, and if it were true, it might have gone a long way toward solving the problem of the "unreasonable effectiveness" of mathematics. In a universe that is *identified* as mathematics, the fact that mathematics fits nature like a glove would hardly be a surprise. Unfortunately, I do not find Tegmark's line of reasoning to be extremely compelling. The leap from the existence of an external reality (independent of humans) to the conclusion that, in Tegmark's words, "You must believe in what I call the mathematical universe hypothesis: that our physical reality is a mathematical structure," involves, in my opinion, a sleight of hand. When Tegmark attempts to characterize what mathematics really is, he says: "To a modern logician, a mathematical structure is precisely this: a set of abstract entities with relations between them." But this modern logician is human! In other words, Tegmark never really *proves* that our mathematics is not invented by humans; he simply assumes it. Furthermore, as the French neurobiologist Jean-Pierre Changeux has pointed out in response to a similar assertion: "To claim physical reality for mathematical objects, on a level of the natural phenomena we study in biology, poses a worrisome epistemological problem it seems to me. How can a physical state, internal to our brain, represent another physical state external to it?"

Most other attempts to place mathematical objects squarely in the external physical reality simply rely on the effectiveness of mathematics in explaining nature as proof. This however assumes that no other explanation for the effectiveness of mathematics is possible, which, as I will show later, is not true.

If mathematics resides neither in the spaceless and timeless Platonic world nor in the physical world, does this mean that mathematics is entirely invented by humans? Absolutely not. In fact, I shall argue in the next section that most of mathematics does consist of discoveries. Before going any further, however, it would be helpful to first examine some of the opinions of contemporary cognitive scientists. The reason is simple—even if mathematics were entirely discovered, these discoveries would still have been made by human mathematicians using their brains.

With the enormous progress in the cognitive sciences in recent years, it was only natural to expect that neurobiologists and psychologists would turn their attention to mathematics, in particular to the search for the foundations of mathematics in human cognition. A cursory glance at the conclusions of most cognitive scientists may initially leave you with the impression that you are witnessing an embodiment of Mark Twain's phrase "To a man with a hammer, everything looks like a nail." With small variations in emphasis, essentially all of the neuropsychologists and biologists determine that mathematics is a human invention. Upon closer examination, however, you find that while the interpretation of the cognitive data is far from being unambiguous, there is no question that the cognitive efforts represent a new and innovative phase in the search for the foundations of mathematics. Here is a small but representative sample of the comments made by the cognitive scientists.

The French neuroscientist Stanislas Dehaene, whose primary interest is in numerical cognition, concluded in his 1997 book *The Number Sense* that "intuition about numbers is thus anchored deep in our brain." This position is in fact close to that of the intuitionists, who wanted to ground all of mathematics in the pure form of intuition of the natural numbers. Dehaene argues that discoveries about the psychology of arithmetic confirm that "number belongs to the 'natural objects of thought,' the innate categories according to which we apprehend the world." Following a separate study conducted with the Mundurukú—an isolated Amazonian indigenous group—Dehaene and his collaborators added in 2006 a similar judgment about geometry: "The spontaneous understanding of geometrical concepts and maps by this remote human community provides evidence that core geometrical knowledge, like basic arithmetic, is a universal constituent of the human mind." Not all cognitive scientists agree with the latter conclusions. Some point out, for instance, that the success of the Mundurukú in the recent geometrical study, in which they had to identify a curve among straight lines, a rectangle among squares, an ellipse among circles, and so on, may have more to do with their visual ability to spot the odd one out, rather than with an innate geometrical knowledge.

The French neurobiologist Jean-Pierre Changeux, who engaged in a fascinating dialogue on the nature of mathematics with the mathematician (of Platonic "persuasion") Alain Connes in *Conservations on Mind, Matter, and Mathematics,* provided the following observation:

> The reason mathematical objects have nothing to do with the sensible world has to do . . . with their generative character, their capacity to give birth to other objects. The point that needs emphasizing here is that there exists in the brain what may be called a "conscious compartment," a sort of physical space for simulation and creation of new objects . . . In certain respects these new mathematical objects are like living beings: like living beings they're physical objects susceptible to very rapid evolution; unlike living beings, with the particular exception of viruses, they evolve in our brain.

Finally, the most categorical statement in the context of invention versus discovery was made by cognitive linguist George Lakoff and psychologist Rafael Núñez in their somewhat controversial book *Where Mathematics Comes From.* As I have noted already in chapter 1, they pronounced:

> Mathematics is a natural part of being human. It arises from our bodies, our brains, and our everyday experiences in the world. [Lakoff and Núñez therefore speak of mathematics as arising from an "embodied mind"] . . . Mathematics is a system of human concepts that makes extraordinary use of the ordinary tools of human cognition . . . Human beings have been responsible for the creation of mathematics, and we remain responsible for maintaining and extending it. The portrait of mathematics has a human face.

The cognitive scientists base their conclusions on what they regard as a compelling body of evidence from the results of numerous experiments. Some of these tests involved functional imaging studies of the brain during the performance of mathematical tasks. Others examined the math competence of infants, of hunter-gatherer groups such as the Mundurukú, who were never exposed to schooling, and of people with various degrees of brain damage. Most of the researchers agree that certain mathematical capacities appear to be innate. For instance, all humans are able to tell at a glance whether they are looking at one, two, or three objects (an ability called *subitizing*). A very limited version of arithmetic, in the form of grouping, pairing, and very simple addition and subtraction, may also be innate, as is perhaps some very basic understanding of geometrical concepts (although this assertion is more controversial). Neuroscientists have also identified regions in the brain, such as the angular gyrus in the left hemisphere, that appear to be crucial for juggling numbers and mathematical computations, but which are not essential for language or the working memory.

According to Lakoff and Núñez, a major tool for advancement beyond these innate abilities is the construction of *conceptual metaphors*—thought processes that translate abstract concepts into more concrete ones. For example, the conception of arithmetic is grounded in the very basic metaphor of object collection. On the other hand, Boole's more abstract algebra of classes metaphorically linked classes to numbers. The elaborate scenario developed by Lakoff and Núñez offers interesting insights into why humans find some mathematical concepts much more difficult than others. Other researchers, such as cognitive neuroscientist Rosemary Varley of the University of Sheffield, suggest that at least some mathematical structures are parasitic on the language faculty—mathematical insights develop by borrowing mind tools used for building language.

The cognitive scientists make a fairly strong case for an association of our mathematics with the human mind, and against Platonism. Interestingly, though, what I regard as possibly the strongest argument against Platonism comes not from neurobiologists, but rather from Sir Michael Atiyah, one of the greatest mathematicians of the twentieth century. I did, in fact, mention his line of reasoning briefly in chapter 1, but I would now like to present it in more detail.

If you had to choose one concept of our mathematics that has the highest probability of having an existence independent of the human mind, which one would you select? Most people would probably conclude that this has to be the natural numbers. What can be more "natural" than 1, 2, 3, . . . ? Even the German mathematician of intuitionist inclinations Leopold Kronecker (1823–91) famously declared: "God created the natural numbers, all else is the work of man." So if one could show that even the natural numbers, as a concept, have their origin in the human mind, this would be a powerful argument in favor of the "invention" paradigm. Here, again, is how Atiyah argues the case: "Let us imagine that intelligence had resided, not in mankind, but in some vast solitary and isolated jelly-fish, buried deep in the depths of the Pacific Ocean. It would have no experience of individual objects, only with the surrounding water. Motion, temperature and pressure would provide its basic sensory data. In such a pure continuum the discrete would not arise and there would be nothing to count." In other words, Atiyah is convinced that even a concept as basic as that of the natural numbers was *created* by humans, by abstracting (the cognitive scientists would say, "through grounding metaphors") elements of the physical world. Put differently, the number 12, for instance, represents an abstraction of a property that is common to all things that come in dozens, in the same way that the word "thoughts" represents a variety of processes occurring in our brains.

The reader might object to the use of the hypothetical universe of the jellyfish to prove this point. He or she may argue that there is only one, inevitable universe, and that every supposition should be examined in the context of this universe. However, this would be tantamount to conceding that the concept of the natural numbers is in fact somehow dependent on the universe of human experiences! Note that this is precisely what Lakoff and Núñez mean when they refer to mathematics as being "embodied."

I have just argued that the concepts of our mathematics originate in the human mind. You may wonder then why I had insisted earlier that much of mathematics is in fact discovered, a position that appears to be closer to that of the Platonists.

Invention *and* Discovery

In our everyday language the distinction between discovery and invention is sometimes crystal clear, sometimes a bit fuzzier. No one would say that Shakespeare discovered Hamlet, or that Madame Curie invented radium. At the same time, new drugs for certain types of diseases are normally announced as discoveries, even though they often involve the meticulous synthesis of new chemical compounds. I would like to describe in some detail a very specific example in mathematics, which I believe will not only help clarify the distinction between invention and discovery but also yield valuable insights into the process by which mathematics evolves and progresses.

In book VI of *The Elements,* Euclid's monumental work on geometry, we find a definition of a certain division of a line into two unequal parts (an earlier definition, in terms of areas, appears in book II). According to Euclid, if a line AB is divided by a point C (figure 62) in such a way that the ratio of the lengths of the two segments (AC/CB) is equal to the whole line divided by the longer segment (AB/AC), then the line is said to have been divided in "extreme and mean ratio." In other words, if $AC/CB = AB/AC$, then each one of these ratios is called the "extreme and mean ratio." Since the nineteenth century, this ratio is popularly known as the *golden ratio.* Some easy algebra can show that the golden ratio is equal to

$$(1 + \sqrt{5}) / 2 = 1.6180339887 \ldots$$

The first question you may ask is why did Euclid even bother to define this particular line division and to give the ratio a name? After all, there are infinitely many ways in which a line could be divided. The answer to this question can be found in the cultural, mystical heritage of the Pythagoreans and Plato. Recall that the Pythagoreans were obsessed with numbers. They thought of the odd numbers

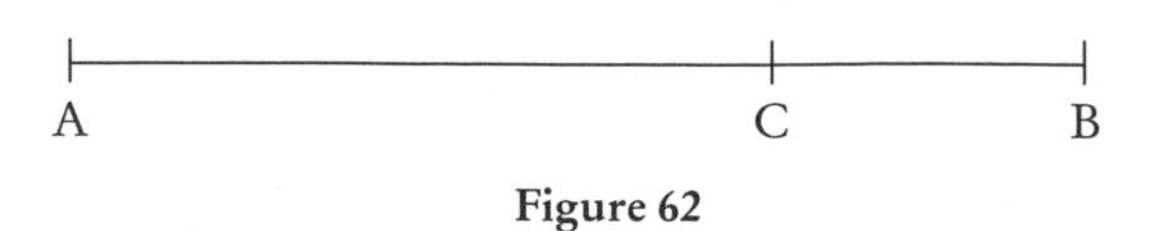

Figure 62

as being masculine and good, and, rather prejudicially, of the even numbers as being feminine and bad. They had a particular affinity for the number 5, the union of 2 and 3, the first even (female) and first odd (masculine) numbers. (The number 1 was not considered to be a number, but rather the generator of all numbers.) To the Pythagoreans, therefore, the number 5 portrayed love and marriage, and they used the pentagram—the five-pointed star (figure 63)—as the symbol of their brotherhood. Here is where the golden ratio makes its first appearance. If you take a regular pentagram, the ratio of the side of any one of the triangles to its implied base (*a*/*b* in figure 63) is precisely equal to the golden ratio. Similarly, the ratio of any diagonal of a regular pentagon to its side (c/d in figure 64) is also equal to the golden ratio. In fact, to construct a pentagon using a straight edge and a compass (the common geometrical construction process of the ancient Greeks) requires dividing a line into the golden ratio.

Plato added another dimension to the mythical meaning of the golden ratio. The ancient Greeks believed that everything in the universe is composed of four elements: earth, fire, air, and water. In *Timaeus,* Plato attempted to explain the structure of matter using the five regular solids that now bear his name—the *Platonic solids* (figure 65). These convex solids, which include the tetrahedron, the cube, the octahedron, the dodecahedron, and the icosahedron, are the only ones in which all the faces (of each individual solid) are the same, and are regular polygons, and where all the vertices of each solid lie on a sphere. Plato associated each of four of the Platonic solids with one of

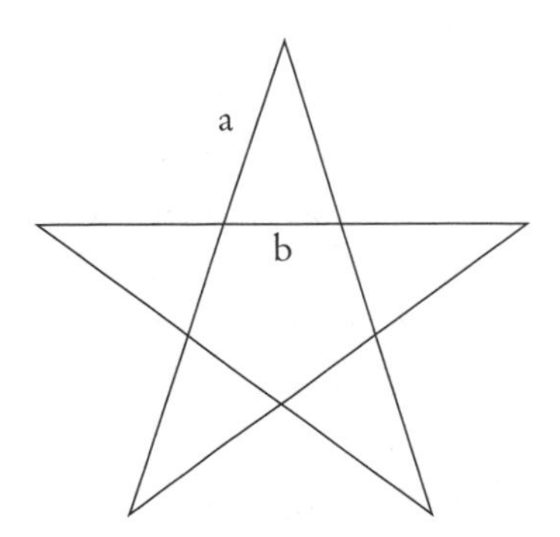

Figure 63

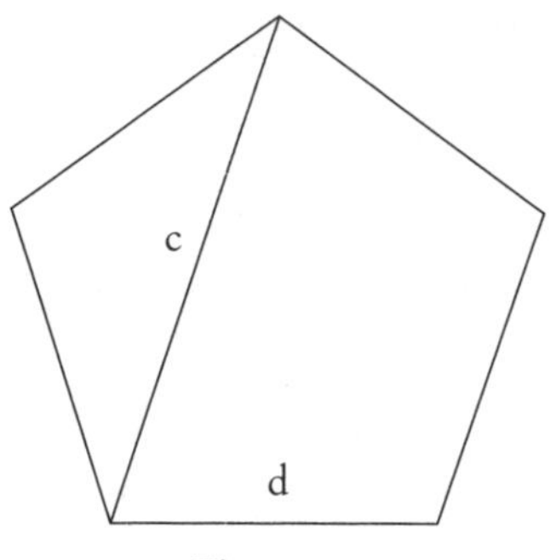

Figure 64

the four basic cosmic elements. For instance, the Earth was associated with the stable cube, the penetrating fire with the pointy tetrahedron, air with the octahedron, and water with the icosahedron. Concerning the dodecahedron (Figure 65d), Plato wrote in *Timaeus:* "As there still remained one compound figure, the fifth, God used it for the whole, broidering it with designs." So the dodecahedron represented the universe as a whole. Note, however, that the dodecahedron, with its twelve pentagonal surfaces, has the golden ratio written all over it. Both its volume and its surface area can be expressed as simple functions of the golden ratio (the same is true for the icosahedron).

History therefore shows that by numerous trials and errors, the Pythagoreans and their followers *discovered* ways to construct certain geometrical figures that to them represented important concepts, such as love and the entire cosmos. No wonder, then, that they, and Euclid (who documented this tradition), *invented the concept* of the golden ratio that was involved in these constructions, and gave it a name. Unlike any other arbitrary ratio, the number 1.618 . . . now became the focus of an intense and rich history of investigation, and it continues to pop up even today in the most unexpected places. For instance, two millennia after Euclid, the German astronomer Johannes Kepler *discovered* that this number appears, miraculously as it were, in relation to a series of numbers known as the *Fibonacci sequence.* The Fibonacci sequence: 1, 1, 2, 3, 5, 8, 13, 21, 34, 55, 89, 144, 233, . . . is characterized by the fact that, starting with the third, each number is the sum of the previous two (e.g., $2 = 1 + 1$; $3 = 1 + 2$;

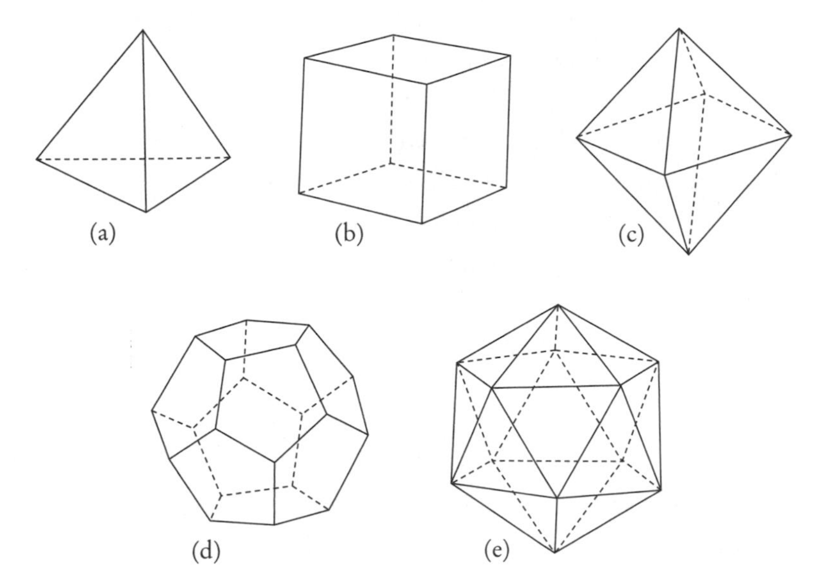

Figure 65

5 = 2 + 3; and so on). If you divide each number in the sequence by the one immediately preceding it (e.g., 144 ÷ 89; 233 ÷ 144; . . .), you find that the ratios oscillate about, but come closer and closer to the golden ratio the farther you go in the sequence. For example, one obtains the following results, rounding the numbers to the sixth decimal place): 144 ÷ 89 = 1.617978; 233 ÷ 144 = 1.618056; 377 ÷ 233 = 1.618026, and so on.

In more modern times, the Fibonacci sequence, and concomitantly the golden ratio, were found to figure in the leaf arrangements of some plants (the phenomenon known as *phyllotaxis*) and in the structure of the crystals of certain aluminum alloys.

Why do I consider Euclid's definition of the concept of the golden ratio an invention? Because Euclid's inventive act singled out this ratio and attracted the attention of mathematicians to it. In China, on the other hand, where the concept of the golden ratio was not invented, the mathematical literature contains essentially no reference to it. In India, where again the concept was not invented, there are only a few insignificant theorems in trigonometry that peripherally involve this ratio.

There are many other examples that demonstrate that the question "Is mathematics a discovery or an invention?" is ill posed. *Our mathematics is a combination of inventions and discoveries.* The axioms of Euclidean geometry as a *concept* were an invention, just as the rules of chess were an invention. The axioms were also supplemented by a variety of invented concepts, such as triangles, parallelograms, ellipses, the golden ratio, and so on. The theorems of Euclidean geometry, on the other hand, were by and large discoveries; they were the paths linking the different concepts. In some cases, the proofs generated the theorems—mathematicians examined what they could prove and from that they deduced the theorems. In others, as described by Archimedes in *The Method,* they first found the answer to a particular question they were interested in, and then they worked out the proof.

Typically, the concepts were inventions. Prime numbers as a *concept* were an invention, but all the theorems about prime numbers were discoveries. The mathematicians of ancient Babylon, Egypt, and China never invented the concept of prime numbers, in spite of their advanced mathematics. Could we say instead that they just did not "discover" prime numbers? Not any more than we could say that the United Kingdom did not "discover" a single, codified, documentary constitution. Just as a country can survive without a constitution, elaborate mathematics could develop without the concept of prime numbers. And it did!

Do we know why the Greeks invented such concepts as the axioms and prime numbers? We cannot be sure, but we could guess that this was part of their relentless efforts to investigate the most fundamental constituents of the universe. Prime numbers were the basic building blocks of numbers, just as atoms were the building blocks of matter. Similarly, the axioms were the fountain from which all geometrical truths were supposed to flow. The dodecahedron represented the entire cosmos and the golden ratio was the concept that brought that symbol into existence.

This discussion highlights another interesting aspect of mathematics—it is a part of the human culture. Once the Greeks invented the axiomatic method, all the subsequent European mathematics followed suit and adopted the same philosophy and practices. Anthropologist

Leslie A. White (1900–1975) tried once to summarize this cultural facet by noting: "Had Newton been reared in Hottentot [South African tribal] culture he would have calculated like a Hottentot." This cultural complexion of mathematics is most probably responsible for the fact that many mathematical discoveries (e.g., of knot invariants) and even some major inventions (e.g., of calculus) were made simultaneously by several people working independently.

Do You Speak Mathematics?

In a previous section I compared the import of the abstract concept of a number to that of the meaning of a word. Is mathematics then some kind of language? Insights from mathematical logic, on one hand, and from linguistics, on the other, show that to some extent it is. The works of Boole, Frege, Peano, Russell, Whitehead, Gödel, and their modern-day followers (in particular in areas such as philosophical syntax and semantics, and in parallel in linguistics), have demonstrated that grammar and reasoning are intimately related to an algebra of symbolic logic. But why then are there more than 6,500 languages while there is only one mathematics? Actually, all the different languages have many design features in common. For instance, the American linguist Charles F. Hockett (1916–2000) drew attention in the 1960s to the fact that all the languages have built-in devices for acquiring new words and phrases (think of "home page"; "laptop"; "indie flick"; and so on). Similarly, all the human languages allow for abstraction (e.g., "surrealism"; "absence"; "greatness"), for negation (e.g., "not"; "hasn't"), and for hypothetical phrases ("If grandma had wheels she might have been a bus"). Perhaps two of the most important characteristics of all languages are their *open-endedness* and their *stimulus-freedom*. The former property represents the ability to create never-before-heard utterances, and to understand them. For instance, I can easily generate a sentence such as: "You cannot repair the Hoover Dam with chewing gum," and even though you have probably never encountered this sentence before, you have no trouble understanding it. Stimulus-freedom is the power to choose how, or even if, one should respond to a received stimulus. For instance, the answer to the question posed by singer/songwriter Carole King in her song "Will You Still Love Me Tomorrow?" could be any of the following: "I don't know if I'll still be alive tomorrow"; "Absolutely"; "I don't even love you today"; "Not as much as I love my dog"; "This is definitely your best song"; or even "I wonder who will win the Australian Open this year." You will recognize that many of these features (e.g., abstraction; negation; open-endedness; and the ability to evolve) are also characteristic of mathematics.

As I noted before, Lakoff and Núñez emphasize the role of metaphors in mathematics. Cognitive linguists also argue that all human languages use metaphors to express almost everything. Even more importantly perhaps, ever since 1957, the year in which the famous linguist Noam Chomsky published his revolutionary work *Syntactic Structures,* many linguistic endeavors have revolved around the concept of a *universal grammar*—principles that govern all languages. In other words, what appears to be a Tower of Babel of diversity may really hide a surprising structural similarity. In fact, if this were not the case, dictionaries that translate from one language into another might have never worked.

You may still wonder why mathematics is as uniform as it is, both in terms of subject matter and in terms of symbolic notation. The first part of this question is particularly intriguing. Most mathematicians agree that mathematics as we know it has evolved from the basic branches of geometry and arithmetic that were practiced by the ancient Babylonians, Egyptians, and Greeks. However, was it truly inevitable that mathematics would start with these particular disciplines? Computer scientist Stephen Wolfram argued in his massive book *A New Kind of Science* that this was not necessarily the case. In particular, Wolfram showed how starting from simple sets of rules that act as short computer programs (known as *cellular automata*), one could develop a very different type of mathematics. These cellular automata could be used (in principle, at least) as the basic tools for modeling natural phenomena, instead of the differential equations that have dominated science for three centuries. What was it, then, that drove the ancient civilizations toward discovering and inventing our special "brand" of mathematics? I don't really know, but it may have had much to do with the particulars of the human perceptual system. Humans detect and perceive edges, straight lines, and smooth curves very easily. Notice, for instance, with what precision you can determine (just by eye) whether a line is perfectly straight, or how effortlessly you are able to distinguish between a circle and a shape that is slightly elliptical. These perceptual abilities may have strongly shaped the human experience of the world, and may have therefore led to a mathematics that was based on discrete objects (arithmetic) and on geometrical figures (Euclidean geometry).

The uniformity in symbolic notation is probably a result of what one might call the "Microsoft Windows effect": The entire world is using Microsoft's operating system—not because this conformity was inevitable, but because once one operating system started to dominate the computer market, everybody had to adopt it to allow for ease in communication and for availability of products. Similarly, the Western symbolic notation imposed uniformity on the world of mathematics.

Intriguingly, astronomy and astrophysics may still contribute to the "invention and discovery" question in interesting ways. The most recent studies of extrasolar planets indicate that about 5 percent of all stars have at least one giant planet (like Jupiter in our own solar system) revolving around them, and that this fraction remains roughly constant, on the average, all across the Milky Way galaxy. While the precise fraction of *terrestrial* (Earth-like) planets is not yet known, chances are that the galaxy is teeming with billions of such planets. Even if only a small (but nonnegligible) fraction of these "Earths" are in the *habitable zone* (the range of orbits that allows for liquid water on a planet's surface) around their host stars, the probability of life in general, and of intelligent life in particular, developing on the surface of these planets is not zero. If we were to discover another intelligent life form with which we could communicate, we could gain invaluable information about the formalisms developed by this civilization to explain the cosmos. Not only would we make unimaginable progress in the understanding of the origin and evolution of life, but we could even compare our logic to the logical system of potentially more advanced creatures.

On a much more speculative note, some scenarios in cosmology (e.g., one known as *eternal inflation*) predict the possible existence of multiple universes. Some of these universes may not only be characterized by different values of the *constants of nature* (e.g., the strengths of the different forces; the mass ratios of subatomic particles), but even by different laws of nature altogether. Astrophysicist Max Tegmark argues that there should even be a universe corresponding to (or that *is,* in his language) each possible mathematical structure. If this were true, this would be an extreme version of the "universe *is* mathematics" perspective—there isn't just one world that can be identified with mathematics, but an entire ensemble of them. Unfortunately, not only is this speculation radical and currently untestable, it also appears (at least in its simplest form) to contradict what has become known as the *principle of mediocrity.* As I have described in chapter 5, if you pick a person at random on the street, you have a 95 percent chance that his or her height would be within two standard deviations from the mean height. A similar argument should apply to the properties of universes. But the number of possible mathematical structures increases dramatically with increasing complexity. This means that the most "mediocre" structure (close to the mean) should be incredibly intricate. This appears to be at odds with the relative simplicity of our mathematics and our theories of the universe, thus violating the natural expectation that our universe should be typical.

Wigner's Enigma

"Is mathematics created or discovered?" is the wrong question to ask because it implies that the answer has to be one or the other and that the two possibilities are mutually exclusive. Instead, I suggest that mathematics is partly created and partly discovered. Humans commonly invent mathematical concepts and discover the relations among those concepts. Some empirical discoveries surely preceded the formulation of concepts, but the concepts themselves undoubtedly provided an incentive for more theorems to be discovered. I should also note that some philosophers of mathematics, such as the American Hilary Putnam, adopt an intermediate position known as *realism*—they believe in the objectivity of mathematical discourse (that is, sentences are true or false, and what makes them true or false is external to humans) without committing themselves, like the Platonists, to the existence of "mathematical objects." Do any of these insights also lead to a satisfactory explanation for Wigner's "unreasonable effectiveness" puzzle?

Let me first briefly review some of the potential solutions proposed by contemporary thinkers. Physics Nobel laureate David Gross writes:

> A point of view that, from my experience, is not uncommon among creative mathematicians—namely that the mathematical structures that they arrive at are not artificial creations of the human mind but rather have a naturalness to them as if they were as real as the structures created by physicists to describe the so-called real world. Mathematicians, in other words, are not inventing new mathematics, they are discovering it. If this is the case then perhaps some of the mysteries that we have been exploring [the "unreasonable effectiveness"] are rendered slightly less mysterious. If mathematics is about structures that are a real part of the natural world, as real as the concepts of theoretical physics, then it is not so surprising that it is an effective tool in analyzing the real world.

In other words, Gross relies here on a version of the "mathematics as a discovery" perspective that is somewhere between the Platonic world and the "universe *is* mathematics" world, but closer to a Platonic viewpoint. As we have seen, however, it is difficult to philosophically support the "mathematics as a discovery" claim. Furthermore, Platonism cannot truly solve the problem of the phenomenal accuracy that I have described in chapter 8, a point acknowledged by Gross.

Sir Michael Atiyah, whose views on the nature of mathematics I have largely adopted, argues as follows:

> If one views the brain in its evolutionary context then the mysterious success of mathematics in the physical sciences is

> at least partially explained. The brain evolved in order to deal with the physical world, so it should not be too surprising that it has developed a language, mathematics, that is well suited for the purpose.

This line of reasoning is very similar to the solutions proposed by the cognitive scientists. Atiyah also recognizes, however, that this explanation hardly addresses the thornier parts of the problem—how does mathematics explain the more esoteric aspects of the physical world. In particular, this explanation leaves the question of what I called the "passive" effectiveness (mathematical concepts finding applications long after their invention) entirely open. Atiyah notes: "The skeptic can point out that the struggle for survival only requires us to cope with physical phenomena at the human scale, yet mathematical theory appears to deal successfully with all scales from the atomic to the galactic." To which his only suggestion is: "Perhaps the explanation lies in the abstract hierarchical nature of mathematics which enables us to move up and down the world scale with comparative ease."

The American mathematician and computer scientist Richard Hamming (1915–98) provided a very extensive and interesting discussion of Wigner's enigma in 1980. First, on the question of the nature of mathematics, he concluded that "mathematics has been made by man and therefore is apt to be altered rather continuously by him." Then, he proposed four potential explanations for the unreasonable effectiveness: (1) selection effects; (2) evolution of the mathematical tools; (3) the limited explanatory power of mathematics; and (4) evolution of humans.

Recall that selection effects are distortions introduced in the results of experiments either by the apparatus being used or by the way in which the data are collected. For instance, if in a test of the efficiency of a dieting program the researcher would reject everyone who drops out of the trial, this would bias the result, since most likely the ones who drop out are those for whom the program wasn't working. In other words, Hamming suggests that at least in some cases, "the original phenomenon arises from the mathematical tools we use and not from the real world . . . a lot of what we see comes from the glasses we put on." As an example, he correctly points out that one can show that any force symmetrically emanating from a point (and conserving energy) in three-dimensional space should behave according to an inverse-square law, and therefore that the applicability of Newton's law of gravity should not be surprising. Hamming's point is well taken, but selection effects can hardly explain the fantastic accuracy of some theories.

Hamming's second potential solution relies on the fact that humans select, and continuously improve the mathematics, to fit a given situation. In other words, Hamming proposes that we are witnessing what we might call an "evolution and natural selection" of mathematical ideas—humans invent a large number of mathematical concepts, and only those that fit are chosen. For years I also used to believe that this was the complete explanation. A similar interpretation was proposed by physics Nobel laureate Steven Weinberg in his book *Dreams of a Final Theory.* Can this be *the* explanation to Wigner's enigma? There is no doubt that such selection and evolution indeed occur. After sifting through a variety of mathematical formalisms and tools, scientists retain those that work, and they do not hesitate to upgrade them or change them as better ones become available. But even if we accept this idea, why are there mathematical theories that can explain the universe at all?

Hamming's third point is that our impression of the effectiveness of mathematics may, in fact, be an illusion, since there is much in the world around us that mathematics does not really explain. In support of this perspective I could note, for instance, that the mathematician Israïl Moseevich Gelfand was once quoted as having said: "There is only one thing which is more unreasonable than the unreasonable effectiveness of mathematics in physics, and this is the unreasonable *ineffectiveness* [emphasis added] of mathematics in biology." I don't think that this in itself can explain away Wigner's problem. It is true that unlike in *The Hitchhiker's Guide to the Galaxy,* we cannot say that the answer to life, the universe, and everything is forty-two. Nevertheless, there is a sufficiently large number of phenomena that mathematics *does* elucidate to warrant an explanation. Moreover, the range of facts and processes that can be interpreted by mathematics continually widens.

Hamming's fourth explanation is very similar to the one suggested by Atiyah—that "Darwinian evolution would naturally select for survival those competing forms of life which had the best models of reality in their minds—'best' meaning best for surviving and propagating."

Computer scientist Jef Raskin (1943–2005), who started the Macintosh project for Apple Computer, also held related views, with a particular emphasis on the role of logic. Raskin concluded that

> human logic was forced on us by the physical world and is therefore consistent with it. Mathematics derives from logic. This is why mathematics is consistent with the physical world. There is no mystery here—though we should not lose our sense of wonder and amazement at the nature of things even as we come to understand them better.

Hamming was less convinced, even by the strength of his own argument. He pointed out that

> if you pick 4,000 years for the age of science, generally, then you get an upper bound of 200 generations. Considering the effects of evolution we are looking for via selection of small chance variations, it does not seem to me that evolution can explain more than a small part of the unreasonable effectiveness of mathematics.

Raskin insisted that "the groundwork for mathematics had been laid down long before in our ancestors, probably over millions of generations." I must say, however, that I do not find this argument particularly convincing. Even if logic had been deeply embedded in our ancestors' brains, it is difficult to see how this ability could have led to abstract mathematical theories of the subatomic world, such as quantum mechanics, that display stupendous accuracy.

Remarkably, Hamming concluded his article with an admission that "all of the explanations I have given when added together simply are not enough to explain what I set out to account for" (namely, the unreasonable effectiveness of mathematics).

So, should we close by conceding that the effectiveness of mathematics remains as mysterious as it was when we started?

Before giving up, let us try to distill the essence of Wigner's puzzle by examining what is known as the *scientific method.* Scientists first learn facts about nature through a series of experiments and observations. Those facts are initially used to develop some sort of qualitative models of the phenomena (e.g., the Earth attracts apples; colliding subatomic particles can produce other particles; the universe is expanding; and so on). In many branches of science even the emerging theories may remain nonmathematical. One of the best examples of a powerfully explanatory theory of this type is Darwin's theory of evolution. Even though natural selection is not based on a mathematical formalism, its success in clarifying the origin of species has been remarkable. In fundamental physics, on the other hand, usually the next step involves attempts to construct mathematical, quantitative theories (e.g., general relativity; quantum electrodynamics; string theory; and so on). Finally, the researchers use those mathematical models to predict new phenomena, new particles, and results of never-before-performed experiments and observations. What puzzled Wigner and Einstein was the incredible success of the last two processes. How is it possible that time after time physicists are able to find mathematical tools that not only explain the existing experimental and observational results, but which also lead to entirely new discernments and new predictions?

I attempt to answer this version of the question by borrowing a beautiful example from mathematician Reuben Hersh. Hersh proposed that in the spirit of the analysis of many such problems in mathematics (and indeed in theoretical physics) one should examine the simplest possible case. Consider the seemingly trivial experiment of putting pebbles into an opaque vase. Suppose you first put in four white pebbles, and later you put in seven black pebbles. At some point in their history, humans learned that for some purposes they could represent a collection of pebbles of any color by an abstract concept that they had invented—a natural number. That is, the collection of white pebbles could be associated with the number 4 (or IIII or IV or whichever symbol was used at the time) and the black pebbles with the number 7. Via experimentation of the type I have described above, humans also discovered that another invented concept—arithmetic addition—represents correctly the physical act of aggregation. In other words, the result of the abstract process denoted symbolically by 4 + 7 can predict unambiguously the final number of pebbles in the vase. What does all of this mean? It means that humans have developed an incredible mathematical tool—one that could reliably predict the result of *any* experiment of this type! This tool is actually much less trivial than it might seem, because the same tool, for instance, does not work for drops of water. If you put four separate drops of water into the vase, followed by seven additional drops, you don't get eleven separate drops of water in the vase. In fact, to make any kind of prediction for similar experiments with liquids (or gases), humans had to invent entirely different concepts (such as weight) and to realize that they have to weigh individually each drop or volume of gas.

The lesson here is clear. The mathematical tools were not chosen arbitrarily, but rather precisely on the basis of their ability to correctly predict the results of the relevant experiments or observations. So at least for this very simple case, their effectiveness was essentially guaranteed. Humans did not have to guess in advance what the correct mathematics would be. Nature afforded them the luxury of trial and error to determine what worked. They also did not have to stick with the same tools for all circumstances. Sometimes the appropriate mathematical formalism for a given problem did not exist, and someone had to invent it (as in the case of Newton inventing calculus, or modern mathematicians inventing various topological/geometric ideas in the context of the current efforts in string theory). In other cases, the formalism had already existed, but someone had to discover that this was a solution awaiting the right problem (as in the case of Einstein using Riemannian geometry, or particle physicists using group theory). The point is that through a burning curiosity, stubborn persistence, creative imagination, and fierce determination, humans were able to find the relevant mathematical formalisms for modeling a large number of physical phenomena.

One characteristic of mathematics that was absolutely crucial for what I dubbed the "passive" effectiveness was its essentially eternal validity. Euclidean geometry remains as correct today as it was in 300 BC. We understand now that its axioms are not inevitable, and rather than representing absolute truths about space, they represent truths within a particular, human-perceived universe and its associated human-invented formalism. Nevertheless, once we comprehend the more limited context, all the theorems hold true. In other words, branches of mathematics get to be incorporated into larger, more comprehensive branches (e.g., Euclidean geometry is only one possible version of geometry), but the correctness within each branch persists. It is this indefinite longevity that has allowed scientists at any given time to search for adequate mathematical tools in the entire arsenal of developed formalisms.

The simple example of the pebbles in the vase still does not address two elements of Wigner's enigma. First, there is the question why in some cases do we seem to get more accuracy out of the theory than we have put into it? In the experiment with the pebbles, the accuracy of the "predicted" results (the aggregation of other numbers of pebbles) is not any better than the accuracy of the experiments that had led to the formulation of the "theory" (arithmetic addition) in the first place. On the other hand, in Newton's theory of gravity, for instance, the accuracy of its predictions proved to far exceed that of the observational results that motivated the theory. Why? A brief re-examination of the history of Newton's theory may provide some insight.

Ptolemy's geocentric model reigned supreme for about fifteen centuries. While the model did not claim any universality—the motion of each planet was treated individually—and there was no mention of physical causes (e.g., forces; acceleration), the agreement with observations was reasonable. Nicolaus Copernicus (1473–1543) published his heliocentric model in 1543, and Galileo put it on solid ground, so to speak. Galileo also established the foundations for the laws of motion. But it was Kepler who deduced from observations the first mathematical (albeit only phenomenological) laws of planetary motion. Kepler used a huge body of data left by the astronomer Tycho Brahe to determine the orbit of Mars. He referred to the ensuing hundreds of sheets of calculations as "my warfare with Mars." Except for two discrepancies, a circular orbit matched all the observations. Still, Kepler was not satisfied with this solution, and he later described his thought process: "If I had believed that we could ignore these eight minutes [of arc; about a quarter of the diameter of a full moon], I would have patched up my hypothesis . . . accordingly. Now, since it was not permissible to disregard, those eight minutes alone pointed the path to a complete reformation in astronomy." The consequences of this meticulousness were dramatic. Kepler inferred that the orbits of the planets are not circular but elliptical, and he formulated two additional, quantitative laws that applied to *all* the planets. When these laws were coupled with Newton's laws of motion, they served as the basis for Newton's law of universal gravitation. Recall, however, that along the way Descartes proposed his theory of vortices, in which planets were carried around the Sun by vortices of circularly moving particles. This theory could not get very far, even before Newton showed it to be inconsistent, because Descartes never developed a systematic mathematical treatment of his vortices.

What do we learn from this concise history? There can be no doubt that Newton's law of gravitation was the work of a genius. But this genius was not operating in a vacuum. Some of the foundations had been painstakingly laid down by previous scientists. As I noted in chapter 4, even much lesser mathematicians than Newton, such as the architect Christopher Wren and the physicist Robert Hooke, independently suggested the inverse square law of attraction. Newton's greatness showed in his unique ability to put it all together in the form of a unifying theory, and in his insistence on providing a mathematical proof of the consequences of his theory. Why was this formalism as accurate as it was? Partly because it treated the most fundamental problem—the forces between two gravitating bodies and the resulting motion. No other complicating factors were involved. It was for this problem and this problem alone that Newton obtained a complete solution. Hence, the fundamental theory was extremely accurate, but its implications had to undergo continuous refinement. The solar system is composed of more than two bodies. When the effects of the other planets are included (still according to the inverse square law), the orbits are no longer simple ellipses. For instance, the Earth's orbit is found to slowly change its orientation in space, in a motion known as *precession*, similar to that exhibited by the axis of a rotating top. In fact, modern studies have shown that, contrary to Laplace's expectations, the orbits of the planets may eventually even become chaotic. Newton's fundamental theory itself, of course, was later subsumed by Einstein's general relativity. And the emergence of that theory also followed a series of false starts and near misses. So the accuracy of a theory cannot be anticipated. The proof of the pudding is in the eating—modifications and amendments continue to be made until the desired accuracy is obtained. Those few cases in which a superior accuracy is achieved in a single step have the appearance of miracles.

There is, clearly, one crucial fact in the background that makes the search for fundamental laws worthwhile. This is the fact that nature has been kind to us by being governed by *universal* laws, rather than by mere parochial bylaws. A hydrogen atom on Earth, at the other edge of the Milky Way galaxy, or even in a galaxy that is ten billion light-years away, behaves in precisely the same manner. And this is true in any direction we look and at any time. Mathematicians and physicists have invented a mathematical term to refer to such properties; they are called *symmetries* and they reflect immunity to changes in location, orientation, or the time you start your clock. If not for these (and other) symmetries, any hope of ever deciphering nature's grand design would have been lost, since experiments would have had to be continuously repeated in every point in space (if life could emerge at all in such a universe). Another feature of the cosmos that lurks in the background of mathematical theories has become known as *locality*. This reflects our ability to construct the "big picture" like a jigsaw puzzle, starting with a description of the most basic interactions among elementary particles.

We now come to the last element in Wigner's puzzle: What is it that guarantees that a mathematical theory should exist at all? In other words, why is there, for instance, a theory of general relativity? Could it not be that there is *no* mathematical theory of gravity?

The answer is actually simpler than you might think. There are indeed no guarantees! There exists a multitude of phenomena for which no precise predictions are possible, even in principle. This category includes, for example, a variety of dynamic systems that develop *chaos,* where the tiniest change in the initial conditions may produce entirely different end results. Phenomena that may exhibit such behavior include the stock market, the weather pattern above the Rocky Mountains, a ball bouncing in a roulette wheel, the smoke rising from a cigarette, and indeed the orbits of the planets in the solar system. This is not to say that mathematicians have not developed ingenious formalisms that can address some important aspects of these problems, but no deterministic predictive theory exists. The entire fields of probability and statistics have been created precisely to tackle those areas in which one does not have a theory that yields much more than what has been put in. Similarly, a concept dubbed *computational complexity* delineates limits to our ability to solve problems by practical algorithms, and Gödel's incompleteness theorems mark certain limitations of mathematics even within itself. So mathematics is indeed extraordinarily effective for some descriptions, especially those dealing with fundamental science, but it cannot describe our universe in all its dimensions. To some extent, scientists have selected what problems to work on based on those problems being amenable to a mathematical treatment.

Have we then solved the mystery of the effectiveness of mathematics once and for all? I have certainly given it my best shot, but I doubt very much that everybody would be utterly convinced by the arguments that I have articulated in this book. I can, however, cite Bertrand Russell in *The Problems of Philosophy*:

> Thus, to sum up our discussion of the value of philosophy; Philosophy is to be studied, not for the sake of any definite answers to its questions, since no definite answers can, as a rule, be known to be true, but rather for the sake of the questions themselves; because these questions enlarge our conception of what is possible, enrich our intellectual imagination and diminish the dogmatic assurance which closes the mind against speculation; but above all because, through the greatness of the universe which philosophy contemplates, the mind is also rendered great, and becomes capable of that union with the universe which constitutes its highest good.

NOTES

Chapter 1. A Mystery

PAGE

1 *As the British physicist James Jeans:* Jeans 1930.

1 *Einstein once wondered:* Einstein 1934.

2 *he singled out geometry as the paradigm:* Hobbes 1651.

2 *Penrose identifies three different:* Penrose beautifully discusses these "three worlds" in *Emperor's New Mind* and *Road to Reality.*

3 *Physics Nobel laureate Eugene Wigner:* Wigner 1960. We shall return to this article many times in this book.

5 *that he emphatically declared:* Hardy 1940.

5 *One of his works was reincarnated:* For a discussion of the Hardy-Weinberg law in context see for example Hedrick 2004.

5 *the British mathematician Clifford Cocks:* Cocks invented in 1973 what has become known as the RSA encryption algorithm, but at the time it was classified. The algorithm was independently invented a few years later by R. Rivest, A. Shamir, and L. Adleman at MIT. See Rivest, Shamir, and Adleman 1978.

5 *to describe all the symmetries of the world:* A popular description of symmetry, group theory, and their intertwined history is given in *The Equation That Couldn't Be Solved* (Livio 2005), Stewart 2007, Ronan 2006, and Du Sautoy 2008.

6 *He noticed that a sequence of numbers:* A wonderful popular description of the emergence of chaos theory can be found in Gleick 1987.

7 *Black-Scholes option pricing formula:* Black and Scholes 1973.

8 *The traveling salesman problem was solved:* A superb but technical description of the problem and its solutions can be found in Applegate et al. 2007.

9 *expressed his views very clearly:* Changeux and Connes 1995.

10 *He once wittily remarked:* Gardner 2003.

10 *While reviewing a book:* Atiyah 1995.

12 *In the words of the French neuroscientist:* Changeux and Connes 1995.

12 *In one place she complains:* A brief biography of Marjory Fleming can be found, for instance, at Wallechinsky and Wallace 1975–81.
13 *author Ian Stewart once put it:* Stewart 2004.

Chapter 2. Mystics: The Numerologist and the Philosopher

PAGE

14 *Descartes was one of the principal architects:* A more detailed description of Descartes' contributions is presented in chapter 4.
15 *"I recognize no matter":* Descartes 1644.
15 *credited with introducing the words:* Iamblichus ca. 300 ADa, b; discussed in Guthrie 1987.
15 *biographies of Pythagoras from the third century:* Laertius ca. 250 AD; Porphyry ca. 270 AD; Iamblichus ca. 300 ADa, b.
15 *finds it difficult to identify:* Aristotle ca. 350 BC; discussed in Burkert 1972.
16 *The Greek historian Herodotus:* Herodotus 440 BC.
16 *Empedocles (ca. 492–432 BC) added in admiration:* Porphyry ca. 270 AD.
17 *For instance, the* monad: A clear discussion of the Pythagorean perspective can be found in Strohmeier and Westbrook 1999.
17 *The English historian of philosophy:* Stanley 1687.
17 *The fact that someone would find numbers:* For a fascinating compilation of properties of numbers see Wells 1986.
18 *Pythagoras asks someone to count:* Cited in Heath 1921.
19 *"I swear by the discoverer":* Iamblichus ca. 300 ADa; discussed in Guthrie 1987.
19 *When two similar strings:* Strohmeier and Westbrook 1999; Stanley 1687.
20 *The word "gnomon" (a "marker"):* T. L. Heath gives a detailed discussion of the term and what it meant at different times (Heath 1921). The mathematician Theon of Smyrna (ca. 70–135 AD) used the term in relation to the figurative expression of numbers described in the text in *Mathematics, Useful for Understanding Plato* (Theon of Smyrna ca. 130 AD).
23 *"If we listen to those who wish":* You will notice that in his comment Proclus does not state specifically what he himself believes with respect to the question of whether Pythagoras was the first to formulate the theorem. The story about the ox appears in the writings of Laertius, Porphyry, and the historian Plutarch (ca. 46–120 AD). It is based on verses by Apollodorus. However, the verses only talk about "that famous proposition" without stating which proposition this was. See Laertius ca. 250 AD, Plutarch ca. 75 AD.

24 *These constructions were clearly known:* Renon and Felliozat 1947, van der Waerden 1983.

24 *The basic philosophy expressed by the table:* This cosmology was based on the notion that reality emerges from the fact that Matter (considered indefinite) is shaped by Form (considered the limit).

25 *The book* Philosophy for Dummies: Morris 1999.

25 *The oldest surviving story:* Joost-Gaugier 2006.

26 *From the perspective of the questions:* Good discussions of the Pythagorean contributions and their influence can be found in Huffman 1999, Riedweg 2005, Joost-Gaugier 2006, and Huffman 2006 in the Stanford Encyclopedia of Philosophy.

27 *One of the Pythagoreans:* Fritz 1945.

27 *the recognition of the existence of "countable":* I do not discuss topics such as transfinite numbers and the works of Cantor and Dedekind in the present book. Excellent popular accounts can be found in Aczel 2000, Barrow 2005, Devlin 2000, Rucker 1995, and Wallace 2003.

27 *the philosopher Iamblichus reports:* Iamblichus ca. 300 ADa, b.

28 *to the Pythagoreans, God was not:* See discussion in Netz 2005.

29 *"the safest generalization that can be made":* Whitehead 1929.

29 *Who was this relentless seeker:* The titles of texts about Plato and his ideas can, of course, by themselves fill an entire volume. Here are just a few texts that I found to be very helpful. On Plato in general: Hamilton and Cairns 1961, Havelock 1963, Gosling 1973, Ross 1951, Kraut 1992. On mathematics: Heath 1921, Cherniss 1951, Mueller 1991, Fowler 1999, Herz-Fischler 1998.

30 *According to an oration by the fourth century:* The oration was written in 362 AD, but it did not give any details on the contents of the inscription. The words of the inscription come from a marginal note in a manuscript of Aelius Aristides. The note may have been written by the fourth century orator Sopatros, and it reads (in a translation by Andrew Barker): "There had been inscribed at the front of the School of Plato, 'Let no one who is not a geometer enter.' [That is] in place of 'unfair' or 'unjust': for geometry pursues fairness and justice." The note seems to imply that Plato's inscription replaced "unfair or unjust person" in a sign that was common in sacred places ("Let no unfair or unjust person enter") with the phrase "one who is not a geometer." This story was later repeated by no fewer than five sixth century Alexandrian philosophers, and it eventually made its way into the book *Chiliades,* by the twelfth century polymath Johannes Tzetzes (ca. 1110–80). For a detailed discussion see Fowler 1999.

31 *I was disappointed to discover:* A summary of many unsuccessful archaeological attempts can be found in Glucker 1978.

32 *The first century philosopher and historian:* Discussed in Cherniss 1945, Mekler 1902.
32 *To which the Neoplatonic philosopher:* Cherniss 1945, Proclus ca. 450.
33 *"What we require is that those who take":* Plato ca. 360 BC.
33 *"The science of figures, to a certain degree":* Washington 1788.
33 *is no more real than shadows projected:* An interesting discussion of the allegory can be found in Stewart 1905.
35 *Plato's views formed the basis:* For interesting discussions of Platonism and its place in the philosophy of mathematics, see Tiles 1996, Mueller 1992, White 1992, Russell 1945, Tait 1996. For excellent presentations in popular texts, see Davis and Hersh 1981, Barrow 1992.
36 *mathematics becomes closely associated with the divine:* For a discussion of this topic see Mueller 2005.
36 *He argued that in true astronomy:* Plato's comments on astronomy and planetary motion appear in the *Republic* (Plato ca. 360 BC), in *Timaeus,* and in *Laws.* G. Vlostos and I. Mueller discuss the implications of Plato's position (Vlostos 1975, Mueller 1992).
38 *to help publicize a novel entitled:* The novel is *Uncle Petros and Goldbach's Conjecture,* by A. K. Doxiadis (Doxiadis 2000).
38 *innocent-looking example known as* Catalan's conjecture: For a detailed description see Ribenboim 1994.
39 *Some mathematicians, philosophers, cognitive scientists:* I shall discuss these opinions extensively in chapter 9.
39 *"According to the prophets, the last":* Bell 1940.

Chapter 3. Magicians: The Master and the Heretic

PAGE

41 *"Some existing things are natural":* Aristotle ca. 330 BCa, b; see also Koyré 1978.
42 *Using a clever thought experiment:* Galileo 1589–92.
43 *virtually complete system of logical inference:* This and other logical constructs will be discussed extensively in chapter 7.
44 *When the historian of mathematics:* Bell 1937.
45 *written by one Heracleides:* This is mentioned in commentaries on the *Measurement of a Circle* by the mathematician Eutocius (ca. 480–540 AD); see Heiberg 1910–15.
45 *more interested in the military accomplishments:* Plutarch ca. 75 AD.
46 *Archimedes was born in Syracuse:* His year of birth has been determined based on the *Chiliades,* by the twelfth century Byzantine writer Johannes Tzetzes.

46 *Archimedes spent some time in Alexandria:* Evidence discussed in Dijksterhuis 1957.

47 *This immediately triggered a solution:* The Roman architect Marcus Vitruvius Pollio (first century BC) tells us the story in his treatise *De Architectura.* (See Vitruvius 1st century BC.) He says that Archimedes immersed in water a piece of gold and a piece of silver, both having the same weight as the wreath. He thus found that the wreath displaced more water than the gold but less than the silver. It is easy to show that from the different volumes of water displaced one can calculate the ratio of the weights of the gold and the silver in the wreath. Therefore, contrary to some popular accounts, Archimedes did not need to use the laws of hydrostatics to solve the problem of the wreath.

47 *has been cited by:* In a letter from Thomas Jefferson to M. Correa de Serra in 1814, he wrote: "The good opinion of mankind, like the lever of Archimedes, with the given fulcrum, moves the world." Lord Byron mentions Archimedes' statement in *Don Juan.* JFK used the phrase in a campaign speech, cited in *The New York Times,* on November 3, 1960. Mark Twain used it in an article entitled "Archimedes" in 1887.

49 *Archimedes used an assembly of mirrors:* A group of MIT students attempted to reproduce the burning of a ship with mirrors in October 2005. Some of them also repeated the experiment for the TV show *Myth Busters.* The results were somewhat inconclusive in that while the students were able to achieve a burning area that was self-sustaining, they did not produce a large ignition. A similar experiment performed in Germany in September 2002 did manage to ignite the sail of a ship by using 500 mirrors. A discussion of the burning mirrors can be found on a website by Michael Lahanas.

49 *According to some accounts:* Those precise words from Archimedes are mentioned in the *Chiliades* by Tzetzes; see Dijksterhuis 1957. Plutarch says simply that Archimedes refused to follow the soldier to Marcellus until he had solved the problem in which he was absorbed (Plutarch ca. 75 AD).

49 *As the British mathematician and philosopher:* Whitehead 1911.

50 *Archimedes' opus covers an astonishing range:* A superb book on Archimedes' work is *The Works of Archimedes* (Heath 1897). Other excellent expositions can be found in Dijksterhuis 1957 and Hawking 2005.

51 *"There are some, king Gelon":* Heath 1897.

53 *The story of this discovery:* For a wonderful description of the history of the Palimpsest Project, see Netz and Noel 2007.

54 *Sometime in the tenth century:* Probably in 975 AD.

54 *The scribe Ioannes Myronas:* Netz and Noel 2007.

56 *I was fortunate enough to meet:* Will Noel, who is the director of the project, arranged for a meeting with William Christens-Barry, Roger Easton, and Keith Knox. This team designed the narrow-band imaging system and invented the algorithm used to reveal some of the text. Image-processing techniques have also been developed by researchers Anna Tonazzini, Luigi Bedini, and Emanuele Salerno.

57 *"I will send you the proofs":* Dijksterhuis 1957.

58 *his anticipation of* integral and differential calculus: For a beautiful description of the history and meaning of calculus see Berlinski 1996.

60 *The Greek mathematician Geminus:* Heath 1921.

60 *he requested it be engraved:* Plutarch ca. 75 AD.

60 *Here is Cicero's rather moving description:* Cicero 1st century BC. For a scholarly analysis of Cicero's text in terms of structure, rhetoric, and symbolic function, see Jaeger 2002.

61 *Galileo Galilei was born in Pisa:* An authoritative modern biography is S. Drake's *Galileo at Work* (Drake 1978). A more popular account is J. Reston's *Galileo: A Life* (Reston 1994). See also Van Helden and Burr 1995. The complete works of Galileo appear (in Italian) in Favaro 1890–1909.

62 *"Those who read his works":* In *The Little Balance,* Galilei 1586.

63 *"wood moves more swiftly":* Galileo 1589–92 (Galilei 1600a and Galilei 1600b). C. B. Schmitt suggests (Schmitt 1969, after D. A. Maklich) that Galileo's statement may be the result of the hand holding a lead ball being more tired than the hand holding a wooden ball, and consequently that the release of the wooden ball is more prompt. An excellent presentation of Galileo's correct ideas on falling bodies can be found in Frova and Marenzana 1998 (McManus's 2006 translation). A superb discussion of Galileo's physics can be found in Koyré 1978.

63 *Viviani created the popular image:* A thorough discussion of Galileo's methods and thought process can be found in Shea 1972, and in Machamer 1998.

63 *"was ignorant not only":* Galileo 1589–92. Galileo profusely criticizes Aristotle in *De Motu.* See Galilei 1600a, b.

63 *Virginia, Livia, and Vincenzio:* The life story of Virginia, later known as Sister Maria Celeste, is beautifully told in Dava Sobel's *Galileo's Daughter* (Sobel 1999).

65 *"About 10 months ago":* Galilei 1610a, b. An excellent description of the work that led to the telescope can be found in is Reeves 2008.

66 *As the historian of science Noel Swerdlow:* Swerdlow 1998. For a detailed description of Galileo's discoveries with the telescope, see Shea 1972, Drake 1990.

66 *Turning his telescope to the Moon:* A more popular and very engaging description of Galileo's discoveries, as well as a general history of the telescope, can be found in Panek 1998.

67 *The importance of the discovery:* Galileo's Copernicanism is discussed extensively by Shea 1998 and Swerdlow 1998.

69 *a playful Galileo sent Kepler:* The letter itself was written to the Tuscan ambassador to Prague, but Galileo enclosed the anagram for Kepler.

69 *Kepler tried unsuccessfully to decipher:* In fact, he wrote to Galileo: "I abjure you not to leave us long in doubt of the meaning. For you see you are dealing with real Germans. Think in what distress you place me by your silence." Quoted in Caspar 1993.

69 *Scheiner argued that it was impossible:* The entire episode is discussed in detail in Shea 1972.

71 *The Scottish poet Thomas Seggett:* The epigram was in Latin. Seggett (1570–1627) had been a pupil with Galileo in Padua. The epigram appears in Favaro's *Le Opere.* A beautiful discussion of poetry related to telescopes can be found in Nicolson's *Modern Philology* (Nicolson 1935).

71 *Sir Henry Wotton, an English diplomat:* Curzon 2004.

72 *Here is the Aristotelian Giorgio Coresio:* Coresio 1612. Also cited in Shea 1972.

72 *the Pisan philosopher Vincenzo di Grazia:* Appears in di Grazia's *Considerazioni* (1612), which is reprinted in Favaro's *Opere di Galileo,* vol. 4, p. 385.

73 *In the draft of his treatise:* Cited in Shea 1972.

74 *His premise of celestial immutability:* The entire story of the controversy over the nature of sunspots is described beautifully in Van Helden 1996 and in Swerdlow 1998. See also Shea 1972.

75 *The entire story of* The Assayer: Galilei 1623.

75 *were delivered by Galileo's disciple:* Antonio Favaro, who edited all of Galileo's works, found that large parts of Guiducci's manuscript (containing the lectures) were written in Galileo's handwriting.

75 *"Let it be granted that my master":* Grassi 1619.

76 *"I believe Sarsi is firmly convinced":* Galilei 1623.

77 Discourses and Mathematical Demonstrations: Galilei 1638.

79 *what was truly at the heart:* Excellent discussions of Galileo's opinions on the relation between science and scripture can be found in Feldberg 1995 and in McMullin 1998.

80 *In a long letter to Castelli:* Appears also in von Gebler 1879.

80 *was clearly at odds with that:* Theologian Melchor Cano stated in 1585 that "not only the words but even every comma [in the scripture] has been supplied by the Holy Spirit." Cited in Vawter 1972.

81 *Galileo's further attempts to rely:* An extensive description can be found in Redondi 1998.
82 Dialogue Concerning the Two Chief: Galilei 1632.
82 *"We condemn you to the formal":* de Santillana 1955.
83 *"Therefore, desiring to remove from":* de Santillana 1955.
84 *"The fact that the Pope":* Beltrán Mari 1994. See also discussion in Frova and Marenzana 1998.

Chapter 4. Magicians: The Skeptic and the Giant

PAGE

86 *"the greatest single step ever made":* Cited in Sedgwick and Tyler 1917.
86 *René Descartes was born on March 31:* There are numerous biographies of Descartes. The classic is Baillet 1691. Other books I found helpful were Vrooman 1970 and the relatively recent Rodis-Lewis 1998. Bell 1937 gives a brief but beautiful summary. Very interesting also are Finkel 1898, Watson 2002, and Grayling 2005.
87 *Descartes asked the first passer-by:* While there is no doubt that Descartes did indeed meet Beeckman on that day, Beeckman never mentions any problem on a billboard in his journal. Beeckman rather says that Descartes "made every effort to prove that in reality the angle does not exist."
87 *whose influence on Descartes' physico-mathematical*: See Gaukroger 2002 for a description.
87 *Descartes experienced three dreams:* Most biographers locate this night as occurring in the town of Ulm in the state of Neuburg. Descartes himself told the story in a notebook that was seen by his early biographers. Only a few transcribed passages have survived. Descartes repeated impressions of these dreams in his *Discourse* (Adam and Tannery 1897–1910). A quite comprehensive description of the dreams and their possible interpretations can be found in Grayling 2005 and Cole 1992.
90 *as Descartes wrote to his friend:* Letter to Pierre Chanut, France's ambassador to Sweden, who was also an amateur philosopher. Adam and Tannery 1897–1910.
90 *Descartes was buried in Sweden:* Originally he was buried in the cemetery of Nord-Malmoe. When the remains were transferred to France, there were rumors (Adam and Tannery 1897–1910) that part of them, the skull in particular, remained in Sweden. In France, the remains were first buried in the Abbey of Sainte-Geneviève, then in the convent of the Petits-Augustines. Finally, the remains were put in the

Saint-Germain-des-Prés Cathedral, in what is today the Saint-Benoit Chapel. I had a hard time finding it, because I couldn't believe that Descartes was not buried all by himself. In fact, in the same chapel are buried the two Benedictines Mabillon and Montfaucon, and there is only the bust of Mabillon.

91 *What makes Descartes a true modern:* For one perspective see Balz 1952.

92 *Descartes recognized that the methods:* The standard, authoritative compilation of Descartes' works is that by Adam and Tannery 1897–1910. Most of my quotes come from this source. Many translations exist of a number of individual works, such as Veitch's 1901 *The Philosophy of Descartes,* which contains *Discourse on Method,* the *Meditations,* and the *Principles of Philosophy.* On Descartes' philosophy of science see also Clarke 1992.

92 *this deluge of troubling doubts:* An excellent introduction to Descartes' philosophy in general can be found in Cottingham 1986. For a discussion of the Cartesian doubt and the ensuing *Cogito* see Wolterstorff 1999, Ricoeur 1996, Sorell 2005, Curley 1993, and Beyssade 1993.

94 *He outlined it in a 106-page appendix:* Descartes 1637. One of the translations of the entire book is P. J. Olscamp's 1965 edition (Descartes 1637a). A beautiful translation of *The Geometry,* which also includes a facsimile of the first edition, is *The Geometry of René Descartes* (translated by D. E. Smith and M. L. Latham; Descartes 1637b).

95 *Descartes discovered a way to represent:* Descartes' mathematical achievements are nicely summarized in Rouse Ball 1908. A beautiful popular description of Descartes' life and work can be found in Aczel 2005. The level of abstraction exhibited in Descartes' algebra is analyzed in Gaukroger 1992.

97 *writing his treatise on cosmology and physics:* Descartes' conviction in the existence of "laws of nature" can be gleaned from a letter he wrote to Mersenne in May 1632: "Now I have become bold enough to seek the cause of the position of each fixed star. For although their distribution seems irregular, in various parts of the universe, I have no doubt that there is between them a natural order which is regular and determinate."

98 *Two of his laws closely resemble:* Adam and Tannery 1897–1910. See also Miller and Miller 1983. A good discussion of Descartes' physics can be found in Garber 1992. A more general description of Descartes' natural philosophy appears in Keeling 1968.

100 *Newton's tomb inside Westminster Abbey:* The monument was erected

in 1731. It was commissioned from William Kent and the Flemish sculptor Michael Rysbrack. In addition to Newton's figure, whose elbow rests on some of his works, the sculpture shows youths carrying emblems of Newton's main discoveries. Behind the sarcophagus there is a pyramid, from the middle of which rises a globe on which several constellations are drawn, as well as the path of the comet of 1681.

101 *Actually, Newton may have written that phrase:* It is impossible to know for sure whether Newton meant this as an insult or not. R.K. Merton did find "on the shoulders of giants" to be a fairly common phrase by Newton's time (Merton 1993).

102 *In his reply to Hooke's letter:* Impressively, Newton's entire correspondence has been collected in Turnbull, Scott, Hall, and Tilling 1959–77.

102 *The feud between the two scientists:* The feud is described in great detail in a few excellent biographies of Newton, including Westfall 1983, Hall 1992, and Gleick 2003.

103 *looked like nothing but a collection:* In an essay published in 1674, Hooke wrote about gravity that its "attractive powers are so much more powerful in operating, by how much nearer the body wrought upon is to their own centers." Hence, while he had the correct intuition, he failed to describe it mathematically.

103 *"We offer this work as the mathematical principles":* There are a number of excellent translations of Newton's *Principia,* including Motte 1729 and Cohen and Whitman 1999 (see Newton 1729). The most accessible with helpful notes is Chandrasekhar's 1995 edited version. The general concept of a law of gravity and its history is discussed extensively in Girifalco 2008, Greene 2004, Hawking 2007, and Penrose 2004.

104 *even in his more experimentally based book on light:* Newton 1730.

105 *In his* Memoirs of Sir Isaac Newton's Life: Stukeley 1752. In addition to the full biographies, there are small books describing certain episodes in the life of Newton or his relatives. Among these I note De Morgan 1885 and Craig 1946.

105 *Irrespective of whether the mythical event:* In his biography of Newton, David Brewster wrote in 1831: "The celebrated apple tree, the fall of one of the apples of which is said to have turned the attention of Newton to the subject of gravity, was destroyed by wind about four years ago; but Mr. Turnor [the proprietor of Newton's house in Woolsthorpe] has preserved it in the form of a chair." Brewster 1831.

106 *It may have all started in Newton's youth:* A good description of Newton's studies of mathematics is given in Hall 1992.

107 *"And the same year [1666] I began":* The memorandum is in the Portsmouth Collection. There are other documents suggesting that Newton did indeed think of the inverse square law of gravity during the plague years. See Whiston 1753, for example.

107 *For reasons that are not entirely clear:* For a general discussion of the reasons for the delay in Newton's announcement of the law of gravitation see Cajori 1928 and Cohen 1982. In the next section I summarize what I regard as the two most convincing suggestions as to what the reasons might have been.

108 *"In 1684 Dr Halley came":* De Moivre was recalling here what Newton had described to him.

110 *Some even speculate that he:* This is suggested by Cohen 1982, to name just one source.

111 *In his address at the bicentenary:* Glaisher 1888.

114 *For Newton, the world's very existence:* In *Principia* he says about God: "He is omnipresent not only *virtually* but also *substantially* . . . He is all eye, all ear, all brain, all arm, all force of sensing, of understanding, and of acting." In a manuscript from the early 1700s, purchased at Sotheby's in 1936 and exhibited in Jerusalem in 2007, Newton used the biblical book of Daniel to calculate the date for the apocalypse. In case you are worried, he reached the conclusion that he saw no reason for "its [the world] ending sooner" than 2060.

114 *The validity of the cosmological, teleological:* For excellent recent discussions of the history of these arguments and an assessment of their logical soundness, see Dennett 2006, Dawkins 2006, and Paulos 2008.

116 *This type of logical maneuvering:* See Dennett 2006, Dawkins 2006, Paulos 2008.

Chapter 5. Statisticians and Probabilists: The Science of Uncertainty

PAGE

118 *The branch of mathematics called* calculus: Extremely accessible descriptions of calculus and its applications can be found in Berlinski 1996, Kline 1967, and Bell 1951. Somewhat more technical, but truly excellent is Kline 1972.

119 *were members of the legendary Bernoulli family:* For some of the achievements of this remarkable family, see Maor 1994, Dunham 1994. See also the "Bernoulli-Edition" (in German) on the University of Basel Web page (http://www.ub.unibas.ch/spez/bernoulli.htm). Information on the project in English can be found at http://www.springer.com/cda/content/document/cda_downloaddocument/bernoulli2005web.Pdf?SGWID=0-0-45-169442-0.

119 *known for their bitter intrafamily feuds:* Described in Hellman 2006.

119 *became known as the problem of the* catenary: An excellent description of the problem, and in particular of Huygens's solution, can be found in Bukowski 2008. The solutions of Bernoulli, Leibniz, and Huygens appear in Truesdell 1960.

120 *"You say that my brother proposed":* Quoted in Truesdell 1960.

122 *in his* Philosophical Essay on Probabilities: Laplace 1814 (translated by Truscot and Emory in 1902).

125 *John Graunt (1620–74) was trained:* Excellent descriptions of Graunt's life and work can be found in Hald 1990, Cohen 2006, and Graunt 1662.

128 *Halley's paper, which had the rather long title:* The paper is reprinted in Newman 1956.

130 *Here is how Jakob Bernoulli described:* Quoted in Newman 1956. His work is summarized in Todhunter 1865.

132 *Adolphe Quetelet was born:* Two excellent books on Quetelet and his work are Hankins 1908 and Lottin 1912. Shorter but also informative pieces can be found in Stigler 1997, Krüger 1987, and Cohen 2006.

132 *"Chance, that mysterious, much abused word":* Quetelet 1828.

134 *was in fact a type that nature:* Quetelet wrote in his memoir on the propensity to crime: "If the average man were determined for a nation he would represent the type of that nation; if he could be determined from the ensemble of men, he would represent the type of the entire human race."

136 *The person who first introduced:* For a popular exposition of the work of Galton and Pearson, see Kaplan and Kaplan 2006.

138 *The serious study of probability:* Recently published, entertaining popular accounts of probability, its history, and its uses include Aczel 2004, Kaplan and Kaplan 2006, Connor 2006, Burger and Starbird 2005, and Tabak 2004.

138 *in a letter dated July 29, 1654:* Todhunter 1865, Hald 1990.

139 *The essence of probability theory:* An excellent, popular, brief description of some of the essential principles of probability theory can be found in Kline 1967.

139 *Probability theory provides us with accurate information:* The relevance of probability theory to many real-life situations is beautifully described in Rosenthal 2006.

140 *The person who brought probability:* For an excellent biography, see Orel 1996.

142 *Mendel published his paper:* Mendel 1865. An English translation can be found on the Web page created by R. B. Blumberg at http://www.mendelweb.org.

142 *While some questions related to the accuracy:* See Fisher 1936, for example.

142 *the influential British statistician:* For a brief description of some of his work see Tabak 2004. Fisher wrote an extremely original, non-technical article about the design of experiments entitled "Mathematics of a Lady Tasting Tea" (see Fisher 1956).

145 *in his book* Ars Conjectandi: For a superb translation see Bernoulli 1713b.

146 *He then proceeded to explain:* Reprinted in Newman 1956.

147 *Shaw once wrote an insightful article:* The article "The Vice of Gambling and the Virtue of Insurance" appears in Newman 1956.

149 *In a pamphlet entitled* The Analyst: The pamphlet was written by George Berkeley in 1734. An edited version by David Wilkins is maintained on the Web; see Berkeley 1734.

Chapter 6. Geometers: Future Shock

PAGE

150 *In his famous book* Future Shock: Toffler 1970.

151 *Hume identified "truths":* Hume 1748.

152 *Kant asked not* what *we can know:* According to Kant, one of the fundamental philosophical tasks is to account for the possibility of synthetic *a priori* knowledge of mathematical concepts. Among the many references, I note Höffe 1994 and Kuehn 2001 for the general concepts. A good discussion of the application to mathematics can be found in Trudeau 1987.

152 *"Space is not an empirical":* Kant 1781.

153 *The first four Euclidean axioms:* For a relatively gentle introduction to Euclidean and non-Euclidean geometries, see Greenberg 1974.

153 *the proofs of the first twenty-eight:* Theorems proven without the fifth postulate are discussed in Trudeau 1987.

154 *Some of those endeavors started:* An excellent description of all the attempts that had eventually led to the development of non-Euclidean geometry can be found in Bonola 1955.

155 *The first to publish an entire treatise:* George Bruce Halsted's 1891 translation of Lobachevsky's "Geometrical Researches on the Theory of Parallels" is included in Bonola 1955.

156 *a young Hungarian mathematician, János Bolyai:* For a biography and a description of his work, see Gray 2004. The reason I have not included a picture of *János* Bolyai is that the picture usually used is of doubtful authenticity. Apparently his only relatively reliable portrait is a relief in the façade of the Palace of Culture in Marosvásárhely.

156 *The manuscript was entitled* The Science Absolute of Space: A facsimile of the original (in Latin) and the translation into English by George Bruce Halsted appear in Gray 2004.

158 *There is very little doubt, however:* An excellent description of the entire episode, from the perspective of Gauss's life and work, can be found in Dunnington 1955. A concise but accurate summary of the claims of Lobachevsky and Bolyai for priority is given in Kline 1972. Some of Gauss's correspondence on non-Euclidean geometry is presented in Ewald 1996.

159 *In a brilliant lecture delivered in Göttingen:* An English translation of the lecture, as well as other seminal papers on non-Euclidean geometries, together with illuminating notes, can be found in Pesic 2007.

161 *Poincaré's views were inspired:* Poincaré 1891.

164 *in the first chapter of the* Ars Magna: Cardano 1545.

165 *In another important book,* Treatise of Algebra: Wallis 1685. A concise summary of Wallis's biography and work can be found in Rouse Ball 1908.

165 *Opinions eventually started to change:* A brief summary of the history is given in Cajori 1926.

165 *In an article entitled "Dimension":* This article appeared in Diderot's *Encyclopédie.* Quoted in Archibald 1914.

166 *stating more assertively in 1797:* Lagrange 1797.

166 *Grassmann, one of twelve children:* An excellent biography and description of Grassmann's work (in German) can be found in Petsche 2006. A good brief summary can be found in O'Connor and Robertson 2005.

167 *It is fascinating to follow:* Relatively accessible (but still technical) descriptions of his work in linear algebra can be found in Fearnley-Sander 1979 and 1982.

168 *By the 1860s* n-*dimensional geometry:* A good introductory text is Sommerville 1929.

168 *by the following "declaration of independence":* The text appears in Ewald 1996.

169 *To which algebraist Richard Dedekind:* The text appears in Ewald 1996.

169 *Here is how the French mathematician:* Stieltjer's first letter to Hermite was dated November 8, 1882. The correspondence between the two mathematicians consists of 432 letters. The full correspondence appears in Hermite 1905. I translated the text that appears here.

170 *"Mathematicians have constructed a very large":* The lecture can be found in O'Connor and Robertson 2007.

Chapter 7. Logicians: Thinking About Reasoning

PAGE

172 *The sign outside a barber shop:* The paradox of the village barber is discussed in many books. See Quine 1966, Rescher 2001, and Sorensen 2003, for example.

172 *Here is how Russell himself described:* Russell 1919. This was Russell's more popular exposition of his ideas in logic.

173 *For completeness, I should note:* Brouwer's intuitionist program is summarized nicely by van Stegt 1998. An excellent popular exposition is by Barrow 1992. The debate between formalism and intuitionism is popularly described in Hellman 2006.

174 *"the meaning of a mathematical statement":* Dummett adds that "an individual cannot communicate what he cannot be observed to communicate: if an individual associated with a mathematical symbol or formula some mental content, where the association did not lie in the use he made of the symbol or formula, then he could not convey that content by means of the symbol or formula, for his audience would be unaware of the association and would have no means of becoming aware of it." Dummett 1978.

174 *logic dealt with the relationships:* An extremely accessible introduction to logic can be found in Bennett 2004. More technical, but brilliant, is Quine 1982. A nice summary of the history of logic, by Czeslaw Lejewski, appears in the *Encyclopaedia Britannica,* 15th edition.

176 *De Morgan was an incredibly prolific:* A concise but insightful description of his life and work is given in Ewald 1996.

177 *recounted in* The Mathematical Analysis of Logic: Boole 1847.

178 *George Boole was born:* For a full-length biography see MacHale 1985.

180 *"The design of the following treatise":* Boole 1854.

182 *In spite of the soundness of Boole's conclusion:* Boole concluded that when it comes to the belief in God's existence, the faith-based, nonlogical "feeble steps of an understanding limited in its faculties and its materials of knowledge, are of more avail than the ambitious attempt to arrive at a certainty unattainable on the ground of natural religion."

184 *Frege managed to publish his first revolutionary work:* Frege 1879. This is one of the most important works in the history of logic.

184 *In his* Basic Laws of Arithmetic: Frege 1893, 1903.

184 *Frege's logical axioms generally:* For a general discussion of Frege's ideas and formalism see Resnik 1980, Demopoulos and Clark 2005,

Zalta 2005 and 2007, and Boolos 1985. For an excellent general discussion of mathematical logic, see DeLong 1970.

185 *He then went on to define all the natural numbers:* Frege 1884.

188 *"all of those classes that are not members":* Russell's paradox and its implications and possible remedies are discussed, for instance, in Boolos 1999, Clark 2002, Sainsbury 1988, and Irvine 2003.

188 *the landmark three-volume* Principia Mathematica: Whitehead and Russell 1910. For a popular but illuminating description of *Principia*'s contents, see Russell 1919.

189 *In the* Principia, *Russell and Whitehead:* For the interaction between Russell's and Frege's ideas, see Beaney 2003. For Russell's logicism, see Shapiro 2000 and Godwyn and Irvine 2003.

189 *Russell proposed a* theory of types: An excellent discussion can be found in Urquhart 2003.

190 *Russell's theory of types was viewed:* The theory of types has indeed fallen out of favor with most mathematicians. However, a similar construct has found new applications in computer programming. See Mitchell 1990, for example.

191 *the German mathematician Ernst Zermelo:* See Ewald 1996 for a description of his contributions.

191 *Zermelo's scheme was further augmented:* Translations of the original papers by Zermelo, Fraenkel, and logician Thoralf Skolem can be found in van Heijenoort 1967. For a relatively gentle introduction to sets and the Zermelo-Fraenkel axioms, see Devlin 1993.

191 *the axiom of choice states:* A very detailed discussion of the axiom can be found in Moore 1982.

192 *known as the* continuum hypothesis: Cantor devised a method to compare the cardinality of infinite sets. In particular, he proved that the cardinality of the set of real numbers is larger than the cardinality of the set of the integers. He then formulated the continuum hypothesis, which stated that there is no set with a cardinality that is strictly between those of the integers and the real numbers. When David Hilbert posed his famous problems in mathematics in 1900, the question of whether the continuum hypothesis held true was his first problem. For a relatively recent discussion of this problem, see Woodin 2001a, b.

193 *by the American mathematician Paul Cohen:* He described his work in Cohen 1966.

193 *mathematics proper consisted simply of a collection:* A good description of the Hilbert program can be found in Sieg 1988. An excellent, updated review of the philosophy of mathematics, and a clear summary of the tensions among logicism, formalism, and intuitionism are presented in Shapiro 2000.

194 *"My investigations in the new grounding":* Hilbert delivered this lecture in Leipzig in September 1922. The text can be found in Ewald 1996.

195 *to his formalist followers:* For a good discussion on formalism see Detlefsen 2005.

195 *considered by some to be the greatest:* R. Monk presents a wonderful biography (Monk 1990).

195 *"If I am unclear about the nature":* In Waismann 1979.

195 *Kurt Gödel was born:* A recent biography is Goldstein 2005. The standard biography has been Dawson 1997.

196 *he published his* incompleteness theorems: Excellent books on Gödel's theorems, their meaning, and their connection with other branches of knowledge include Hofstadter 1979, Nagel and Newman 1959, and Franzén 2005.

197 *"But, despite their remoteness":* Gödel 1947.

197 *Gödel the man was every bit:* A comprehensive description of Gödel's philosophical views and how he related philosophical ideas to the foundations of mathematics can be found in Wang 1996.

198 *"It was in 1946 that Gödel":* Morgenstern 1971.

202 *Now all that existed was an incomplete:* This is clearly a huge oversimplification, allowed only in a popular text. In fact, serious attempts in logicism continue even today. These typically assume that many mathematical truths are knowable *a priori.* See Wright 1997 and Tennant 1997, for example.

Chapter 8. Unreasonable Effectiveness?

PAGE

204 *Various knots were even given:* An interesting book on making knots is Ashley 1944.

204 *The mathematical theory of knots:* Vandermonde 1771. An excellent review of the history of knot theory can be found in Przytycki 1992. A lively introduction to the theory itself is presented in Adams 1994. A popular account is given in Neuwirth 1979, Peterson 1988, and Menasco and Rudolph 1995.

205 *Thomson's efforts concentrated on formulating:* Excellent descriptions are presented by Sossinsky 2002 and Atiyah 1990.

206 *Tait started his classification:* Tait 1898, Sossinsky 2002. A brief, well-written biography of Tait can be found in O'Connor and Robertson 2003.

207 *Maxwell offered the following rhyme:* Knott 1911.

208 *University of Nebraska professor:* Little 1899.

208 *Topology—the rubber-sheet geometry:* A technical but still elementary introduction to topology is provided in Messer and Straffin 2006.
209 *the New York lawyer:* Perko 1974.
209 *A breakthrough in knot theory came:* Alexander 1928.
210 *the prolific English-American mathematician:* Conway 1970.
211 *An examination of that relation eventually revealed:* Jones 1985.
213 *in a wide range of sciences:* For instance, mathematician Louis Kauffman has demonstrated a relationship between the Jones polynomial and statistical physics. An excellent but technical book on physics applications is Kauffman 2001.
214 *The agents that take care:* An excellent description of knot theory and the action of enzymes is given in Summers 1995. See also Wasserman and Cozzarelli 1986.
216 *String theory appears to be:* For wonderful popular accounts of string theory, its successes and problems, see Greene 1999, Randall 2005, Krauss 2005, and Smolin 2006. For a technical introduction, see Zweibach 2004.
217 *string theorists Hirosi Ooguri and Cumrun Vafa:* Ooguri and Vafa 2000.
217 *created an unexpected relation:* Witten 1989.
217 *rethought from a purely mathematical perspective:* Atiyah 1989; see Atiyah 1990 for a broader perspective.
218 *Eric Adelberger, Daniel Kapner, and their collaborators:* Kapner et al. 2007.
219 *Einstein had a very strong reason:* There are many excellent expositions of the ideas of special and general relativity. I'll mention here only a few that I have particularly liked: Davies 2001, Deutsch 1997, Ferris 1997, Gott 2001, Greene 2004, Hawking and Penrose 1996, Kaku 2004, Penrose 2004, Rees 1997, and Smolin 2001. A recent, wonderful description of Einstein the man and his ideas is given in Isaacson 2007. Previous superb depictions of Einstein and his world include Bodanis 2000, Lightman 1993, Overbye 2000, and Pais 1982. For a nice collection of original papers, see Hawking 2007.
222 *The most recent test was the result:* Kramer et al. 2006.
223 *a group of physicists at Harvard University:* Odom et al. 2006.
224 *In the late 1960s, physicists:* An excellent description can be found in Weinberg 1993.

Chapter 9. On the Human Mind, Mathematics, and the Universe

PAGE

225 *Here is how mathematicians:* Davis and Hersh 1981.
226 *representing the Platonic point of view:* Hardy 1940.

226 *expressed precisely the opposite perspective:* Kasner and Newman 1989.

228 *Those who believe that mathematics exists:* One of the best popular discussions of the nature of mathematics can be found in Barrow 1992. A slightly more technical but still very accessible review of some of the major ideas is given in Kline 1972.

228 *Since I have already discussed pure Platonism:* For another excellent discussion of many of the topics in the present book, see Barrow 1992.

228 *Tegmark argues that:* Tegmark 2007a, b.

229 *in response to a similar assertion:* Changeux and Connes 1995.

230 *concluded in his 1997 book:* Dehaene 1997.

230 *Dehaene and his collaborators:* Dehaene et al. 2006.

230 *Not all cognitive scientists agree:* See Holden 2006, for example.

231 *provided the following observation:* Changeux and Connes 1995.

231 *the most categorical statement:* Lakoff and Núñez 2000.

232 *Neuroscientists have also identified:* See Ramachandran and Blakeslee 1999, for instance.

232 *cognitive neuroscientist Rosemary Varley:* Varley et al. 2005; Klessinger et al. 2007.

233 *Here, again, is how Atiyah argues:* Atiyah 1995.

234 *Since the nineteenth century:* For a very detailed description of the Golden Ratio, its history and properties, see Livio 2002, and also Herz-Fischler 1998.

238 *Prime numbers as a* concept: A good discussion of these ideas is provided in an article by Yehuda Rav in Hersh 2000.

238 *Anthropologist Leslie A. White:* White 1947.

239 *drew attention in the 1960s to the fact:* For a popularized description see Hockett 1960.

239 *The former property represents the ability:* For a readable discussion of language and the brain see Obler and Gjerlow 1999.

240 *are also characteristic of mathematics:* The similarities between language and mathematics are also discussed by Sarrukai 2005 and Atiyah 1994.

240 *Noam Chomsky published his revolutionary work:* Chomsky 1957. For more on linguistics, an excellent review can be found in Aronoff and Rees-Miller 2001. A popularized, very interesting perspective is given in Pinker 1994.

240 *Computer scientist Stephen Wolfram argued:* Wolfram 2002.

242 *Astrophysicist Max Tegmark argues:* Tegmark identified four distinct types of parallel universes. In "Level I," there are universes with the same laws of physics but different initial conditions. In "Level II,"

there are universes with the same equations of physics but perhaps different constants of nature. "Level III" employs the "many worlds" interpretation of quantum mechanics, and in "Level IV," there are different mathematical structures. Tegmark 2004, 2007b.

242 *to contradict what has become known as the* principle of mediocrity: For an excellent discussion of this topic see Vilenkin 2006.

242 *adopt an intermediate position known as* realism: Putnam 1975.

243 *Let me first briefly review:* There are other opinions that I do not discuss. For instance, Steiner (2005) argues that Wigner does not show that the examples that he gives for "unreasonable effectiveness" have anything to do with the fact that the concepts are mathematical.

243 *Physics Nobel laureate David Gross writes:* Gross 1988. For further discussion of the relationship between mathematics and physics, see Vafa 2000.

243 *Sir Michael Atiyah, whose views on the nature:* Atiyah 1995; see also Atiyah 1993.

244 *mathematician and computer scientist Richard Hamming:* Hamming 1980.

245 *A similar interpretation was proposed:* Weinberg 1993.

245 *Gelfand was once quoted:* In Borovik 2006.

246 *Raskin concluded that:* Raskin 1998.

247 *Hersh proposed that in the spirit:* Excellent article by Hersh in Hersh 2000.

249 *Kepler used a huge body of data:* Kepler's own books, reprinted as Kepler 1981 and 1997, make for very interesting reading in the history of science. Excellent biographies include Caspar 1993 and Gingerich 1973.

251 *the orbits of the planets may eventually:* For a review, see Lecar et al. 2001.

251 *The answer is actually simpler:* An interesting discussion of the utility of mathematics appears in Raymond 2005. Insightful perspectives on Wigner's enigma are also found in Wilczek 2006, 2007.

252 *Bertrand Russell in* The Problems of Philosophy: Russell 1912.

BIBLIOGRAPHY

Aczel, A. D. 2000. *The Mystery of the Aleph: Mathematics, the Kabbalah, and the Search for Infinity* (New York: Four Walls Eight Windows).

——. 2004. *Chance: A Guide to Gambling, Love, the Stock Market, and Just about Everything Else* (New York: Thunder's Mouth Press).

——. 2005. *Descartes' Secret Notebook* (New York: Broadway Books).

Adam, C., and Tannery, P., eds. 1897–1910. *Oeuvres des Descartes.* Revised edition 1964–76 (Paris: Vrin/CNRS). The most comprehensive translation into English is Cottingham, J., Stoothoff, R., and Murdoch, D., eds. 1985. *The Philosophical Writing of Descartes* (Cambridge: Cambridge University Press).

Adams, C. 1994. *The Knot Book: An Elementary Introduction to the Mathematical Theory of Knots* (New York: W. H. Freeman).

Alexander, J. W. 1928. *Transactions of the American Mathematical Society*, 30, 275.

Applegate, D. L., Bixby, R. E., Chvátal, V., and Cook, W. J. 2007. *The Traveling Salesman Problem* (Princeton: Princeton University Press).

Archibald, R. C. 1914. *American Mathematical Society Bulletin*, 20, 409.

Aristotle. Ca. 350 BC. *Metaphysics.* In Barnes, J., ed. 1984. *The Complete Works of Aristotle* (Princeton: Princeton University Press).

——. Ca. 330 BCa. *Physics.* Translated by R. P. Hardie and R. K. Gaye. http://people.bu.edu/wwildman/WeirdWildWeb/courses/wphil/readings/wphil_rdg07_physics_entire.htm (public domain English translation).

——. Ca. 330 BCb. *Physics.* Translated by P. H. Wickstead and F. M. Cornford, 1960 (London: Heinemann).

Aronoff, M., and Rees-Miller, J. 2001. *The Handbook of Linguistics* (Oxford: Blackwell Publishing).

Ashley, C. W. 1944. *The Ashley Book of Knots* (New York: Doubleday).

Atiyah, M. 1989. *Publications Mathématiques de l'Inst. des Hautes Etudes Scientifiques*, Paris, 68, 175.

——. 1990. *The Geometry and Physics of Knots* (Cambridge: Cambridge University Press).

——. 1993. *Proceedings of the American Philosophical Society*, 137(4), 517.

——. 1994. *Supplement to Royal Society News*, 7, (12), (i).

——. 1995. *Times Higher Education Supplement*, 29 September.

Baillet, A. 1691. *La Vie de M. Des-Cartes* (Paris: Daniel Horthemels). Photographic facsimiles were published in 1972 (Hildesheim: Olms) and 1987 (New York: Garner).

Balz, A. G. A. 1952. *Descartes and the Modern Mind* (New Haven: Yale University Press).

Barrow, J. D. 1992. *Pi in the Sky: Counting, Thinking, and Being* (Oxford: Clarendon Press).

——. 2005. *The Infinite Book: A Short Guide to the Boundless, Timeless and Endless* (New York: Pantheon).

Beaney, M. 2003. In Griffin, N., ed. *The Cambridge Companion to Bertrand Russell* (Cambridge: Cambridge University Press).

Bell, E. T. 1937. *Men of Mathematics: The Lives and Achievements of the Great Mathematicians from Zeno to Poincaré* (New York: Touchstone).

——. 1940. *The Development of Mathematics* (New York: McGraw-Hill).

——. 1951. *Mathematics: Queen and Servant of Science* (New York: McGraw-Hill).

Beltrán Mari, A. 1994. "Introduction." In Galilei, G. *Diálogo Sobre los Dos Máximos Sistemas del Mundo* (Madrid: Alianza Editorial).

Bennett, D. 2004. *Logic Made Easy: How to Know When Language Deceives You* (New York: W. W. Norton).

Berkeley, G. 1734. "The Analyst: Or a Discourse Addressed to an Infidel Mathematician, D. R. Wilkins, ed. http:///www.maths.tcd.ie/pub/HistMath/People/Berkeley/Analyst/Analyst.html.

Berlinski, D. 1996. *A Tour of the Calculus* (New York: Pantheon Books).

Bernoulli, J. 1713a. *The Art of Conjecturing* [*Ars Conjectandi*]. Translated by E.D. Sylla, with introduction and notes, 2006 (Baltimore: Johns Hopkins University Press).

——. 1713b. *Ars Conjectandi* (Basel: Tharnisiorum).

Beyssade, M. 1993. "The Cogito." In Voss, S., ed. *Essays on the Philosophy and Science of René Descartes* (Oxford: Oxford University Press).

Black, F., and Scholes, M. 1973. *Journal of Political Economy*, 81(3), 637.

Bodanis, D. 2000. *$E = mc^2$: A Biography of the World's Most Famous Equation* (New York: Walker).

Bonola, R. 1955. *Non-Euclidean Geometry.* Translated by H. S. Carshaw. (New York: Dover Publications). This is a republication of the 1912 translation (Chicago: Open Court Publishing Company).

Boole, G. 1847. *The Mathematical Analysis of Logic, Being an Essay towards a Calculus of Deductive Reasoning.* In Ewald, W. 1996. *From*

Kant to Hilbert: A Source Book in the Foundations of Mathematics (Oxford: Clarendon Press).

——. 1854. *An Investigation of the Laws of Thought on Which Are Founded the Mathematical Theories of Logic and Probabilities* (London: Macmillan). Reprinted 1958 (Mineola, N.Y.: Dover Publications).

Boolos, G. 1985. *Mind*, 94, 331.

——. 1999. *Logic, Logic, Logic* (Cambridge, Mass.: Harvard University Press).

Borovik, A. 2006. *Mathematics under the Microscope.* http://www.maths.manchester.ac.uk/%7Eavb/micromathematics/downloads.

Brewster, D. 1831. *The Life of Sir Isaac Newton* (London: John Murray, Albemarle Street).

Bukowski, J. 2008. *The College Mathematics Journal*, 39(1), 2.

Burger, E. B., and Starbird, M. 2005. *Coincidences, Chaos, and All That Math Jazz: Making Light of Weighty Ideas* (New York: W. W. Norton).

Burkert, W. 1972. *Lore and Science in Ancient Pythagoreanism* (Cambridge, Mass.: Harvard University Press).

Cajori, F. 1926. *The American Mathematical Monthly*, 33(8), 397.

——. 1928. In The History of Science Society. *Sir Isaac Newton 1727–1927: A Bicentenary Evaluation of His Work* (Baltimore: The Williams & Wilkins Company).

Cardano, G. 1545. *Artis Magnae, sive de regulis algebraices.* Published in 1968 under the title *The Great Art or the Rules of Algebra*, translated and edited by T. R. Witmer (Cambridge, Mass.: MIT Press).

Caspar, M. 1993. *Kepler.* Translated by C. D. Hellman (Mineola, N.Y.: Dover Publications).

Chandrasekhar, S. 1995. *Newton's "Principia" for the Common Reader* (Oxford: Clarendon Press).

Changeux, J.-P., and Connes, A. 1995. *Conversations on Mind, Matter, and Mathematics* (Princeton: Princeton University Press).

Cherniss, H. 1945. *The Riddle of the Early Academy* (Berkeley: University of California Press). Reprinted 1980 (New York: Garland).

——. 1951. *Review of Metaphysics*, 4, 395.

Chomsky, N. 1957. *Syntactic Structures* (The Hague: Mouton & Co.).

Cicero. 1st century BC. *Discussion at Tusculam* [sometimes translated as *Tusculan Disputations*]. In Grant, M., trans. 1971. *Cicero: On the Good Life* (London: Penguin Classics).

Clark, M. 2002. *Paradoxes from A to Z* (London: Routledge).

Clarke, D. M. 1992. In Cottingham, J., ed. *The Cambridge Companion to Descartes* (Cambridge: Cambridge University Press).

Cohen, I. B. 1982. In Bechler, Z., ed. *Contemporary Newtonian Research* (Dordrecht: Reidel).

——. 2006. *The Triumph of Numbers* (New York: W. W. Norton & Company).

Cohen, P. J. 1966. *Set Theory and the Continuum Hypothesis* (New York: W. A. Benjamin).

Cole, J. R. 1992. *The Olympian Dreams and Youthful Rebellion of René Descartes* (Champaign: University of Illinois Press).

Connor, J. A. 2006. *Pascal's Wager: The Man Who Played Dice with God* (New York: HarperCollins).

Conway, J. H. 1970. In Leech, J., ed. *Computational Problems in Abstract Algebra* (Oxford: Pergamon Press).

Coresio, G. 1612. *Operetta intorno al galleggiare de' corpi solidi.* Reprinted in Favaro, A. 1968. *Le Opere di Galileo Galilei.* Edizione Nazionale (Florence: Barbera).

Cottingham, J. 1986. *Descartes* (Oxford: Blackwell).

Craig, Sir J. 1946. *Newton at the Mint* (Cambridge: Cambridge University Press).

Curley, E. 1993. In Voss, S., ed. *Essays on the Philosophy and Science of René Descartes* (Oxford: Oxford University Press).

Curzon, G. 2004. *Wotton and His Words: Spying, Science and Venetian Intrigues* (Philadelphia: Xlibris Corporation).

Davies, P. 2001. *How to Build a Time Machine* (New York: Allen Lane).

Davis, P. J., and Hersh, R. 1981. *The Mathematical Experience* (Boston: Birkhaüser). Revised edition 1998 (Boston: Mariner Books).

Dawkins, R. 2006. *The God Delusion* (New York: Houghton Mifflin Company).

Dawson, J. 1997. *Logical Dilemmas: The Life and Work of Kurt Gödel* (Natick, Mass.: A. K. Peters).

Dehaene, S. 1997. *The Number Sense* (Oxford: Oxford University Press).

Dehaene, S., Izard, V., Pica, P., and Spelke, E. 2006. *Science*, 311, 381.

DeLong, H. 1970. *A Profile of Mathematical Logic* (Reading, Mass.: Addison-Wesley). Republished 2004 (Mineola, N.Y.: Dover Publications).

Demopoulos, W., and Clark, P. 2005. In Shapiro, S., ed. *The Oxford Handbook of Philosophy of Mathematics and Logic* (Oxford: Oxford University Press).

De Morgan, A. 1885. *Newton: His Friend: and His Niece* (London: Elliot Stock).

Dennett, D. C. 2006. *Breaking the Spell: Religion as a Natural Phenomenon* (New York: Viking).

De Santillana, G. 1955. *The Crime of Galileo* (Chicago: University of Chicago Press).

Descartes, R. 1637a. *Discourse on Method, Optics, Geometry, and Meteorology.* Translated by P. J. Olscamp, 1965 (Indianapolis: The Bobbs-Merrill Company).

——. 1637b. *The Geometry of René Descartes.* Translated by D. E. Smith and M. L. Latham, 1954 (Mineola, N.Y.: Dover Publications).

——. 1644. *Principles of Philosophy*, II:64. In Cottingham, J., Stoothoff, R., and Murdoch, D., eds. 1985. *Philosophical Works of Descartes* (Cambridge: Cambridge University Press).

——. 1637–1644. *The Philosophy of Descartes: Containing the Method, Meditations, and Other Works.* Translated by J. Veitch, 1901 (New York: Tudor Publishing).

Detlefsen, M. 2005. In Shapiro, S., ed. *The Oxford Handbook of Philosophy of Mathematics and Logic* (Oxford: Oxford University Press).

Deutsch, D. 1997. *The Fabric of Reality* (New York: Allen Lane).

Devlin, K. 1993. *The Joy of Sets: Fundamentals of Contemporary Set Theory*, 2nd ed. (New York: Springer-Verlag).

——. 2000. *The Math Gene: How Mathematical Thinking Evolved and Why Numbers Are like Gossip* (New York: Basic Books).

Dijksterhuis, E. J. 1957. *Archimedes* (New York: The Humanities Press).

Doxiadis, A. K. 2000. *Uncle Petros and Goldbach's Conjecture* (New York: Bloomsbury).

Drake, S. 1978. *Galileo at Work: His Scientific Biography* (Chicago: University of Chicago Press).

——. 1990. *Galileo: Pioneer Scientist* (Toronto: University of Toronto Press).

Dummett, M. 1978. *Truth and Other Enigmas* (Cambridge, Mass.: Harvard University Press).

Dunham, W. 1994. *The Mathematical Universe: An Alphabetical Journey through the Great Proofs, Problems and Personalities* (New York: John Wiley & Sons).

Dunnington, G. W. 1955. *Carl Friedrich Gauss: Titan of Science* (New York: Hafner Publishing).

Du Sautoy, M. 2008. *Symmetry: A Journey into the Patterns of Nature* (New York: Harper Collins).

Einstein, A. 1934. "Geometrie und Erfuhrung." In *Mein Weltbild* (Frankfurt am Main: Ullstein Materialien).

Ewald, W. 1996. *From Kant to Hilbert: A Source Book in the Foundations of Mathematics* (Oxford: Clarendon Press).

Favaro, A., ed. 1890–1909. *Le Opere di Galileo Galilei, Edizione Nationale* (Florence: Barbera). There have been a number of reprints, the most recent 1964–66. This text is searchable online at http://www.imss.fi.it/istituto/index.html.

Fearnley-Sander, D. 1979. *The American Mathematical Monthly*, 86(10), 809.

——. 1982. *The American Mathematical Monthly*, 89(3), 161.

Feldberg, R. 1995. *Galileo and the Church: Political Inquisition or Critical Dialogue* (Cambridge: Cambridge University Press).

Ferris, T. 1997. *The Whole Shebang* (New York: Simon & Schuster).

Finkel, B. F. 1898. "Biography: René Descartes." *American Mathematical Monthly*, 5(8–9), 191.

Fisher, R. A. 1936. *Annals of Science*, 1, 115.

——. 1956. In Newman, J. R., ed. *The World of Mathematics* (New York: Simon & Schuster).

Fowler, D. 1999. *The Mathematics of Plato's Academy* (Oxford: Clarendon Press).

Franzén, T. 2005. *Gödel's Theorem: An Incomplete Guide to Its Use and Abuse* (Wellesley, Mass.: A. K. Peters).

Frege, G. 1879. *Begriffsschrift, eine der arithmetischen nachgebildete Formelsprache des reinen Denkens* (Halle, Germany: L. Nebert). Translated by S. Bauer-Mengelberg in van Heijenoort, J., ed. 1967. *From Frege to Gödel: A Source Book in Mathematical Logic* (Cambridge, Mass.: Harvard University Press).

——. 1884. *Der Grundlagen der Arithmetik* (Breslau: Koebner). Translated, by J. L. Austin, 1974. *The Foundations of Arithmetic* (Oxford: Basil Blackwell).

——. 1893. *Grundgesetze der Arithmetik*, bond I (Jena: Verlag Hermann Pohle). This was partially translated in 1964, in Furth, M., ed. *The Basic Laws of Arithmetic* (Berkeley: University of California Press).

——. 1903. *Grundgesetze der Arithmetik*, bond II (Jena: Verlag Hermann Pohle).

Fritz, K. von. 1945. "The Discovery of Incommensurability by Hipposus of Metapontum." *Annals of Mathematics*, 46, 242.

Frova, A., and Marenzana, M. 1998. *Thus Spoke Galileo: The Great Scientist's Ideas and Their Relevance to the Present Day.* Translated by J. McManus, 2006 (Oxford: Oxford University Press).

Galilei, G. 1586. *The Little Balance.* In *Galileo and the Scientific Revolution.* Translated by L. Fermi and G. Bernardini. (New York: Basic Books). This is a translation of Favaro, A., ed. 1890–1909. *Le Opere di Galileo Galilei* (Florence: G. Barbera).

——. Ca. 1600a. *On Mechanics.* Translated by S. Drake, 1960 (Madison: University of Wisconsin Press).

——. Ca. 1600b. *On Motion.* Translated by I. E. Drabkin, 1960 (Madison: University of Wisconsin Press).

——. 1610a. *Sidereal Nuncius, or The Sidereal Messenger.* Translated by A. Van Helden, 1989. (Chicago: University of Chicago Press).

——. 1610b. *The Sidereal Messenger* [*Sidereus Nuncius*]. In Drake, S. 1983. *Telescopes, Tides and Tactics* (Chicago: University of Chicago Press).

——. 1623. *The Assayer* [*Il Saggiatore*]. In *The Controversy on the Comets of 1618.* Translated by S. Drake and C. D. O'Malley, 1960 (Philadelphia: University of Pennsylvania Press).

——. 1632. *Dialogue Concerning the Two Chief World Systems.* Translated by S. Drake, 1967 (Berkeley: University of California Press).

——. 1638. *Discourses on the Two New Sciences.* Translated by S. Drake, 1974 (Madison: University of Wisconsin Press).

Garber, D. 1992. In Cottingham, J., ed. *The Cambridge Companion to Descartes* (Cambridge: Cambridge University Press).

Gardner, M. 2003. *Are Universes Thicker than Blackberries?* (New York: W. W. Norton).

Gaukroger, S. 1992. In Cottingham, J., ed. *The Cambridge Companion to Descartes* (Cambridge: Cambridge University Press).

——. 2002. *Descartes's System of Natural Philosophy* (Cambridge: Cambridge University Press).

Gingerich, O. 1973. "Kepler, Johannes." In Gillespie, C. C., ed. *Dictionary of Scientific Biography*, vol. 7 (New York: Scribners).

Girifalco, L. A. 2008. *The Universal Force* (Oxford: Oxford University Press).

Glaisher, J. W. L. 1888. Bicentenary Address, *Cambridge Chronicle*, April 20, 1888.

Gleick, J. 1987. *Chaos: Making a New Science* (New York: Viking).

——. 2003. *Isaac Newton* (New York: Vintage Books).

Glucker, J. 1978. *Antiochus and the Late Academy*, hypomnemata 56 (Göttingen: Vandenhoeck & Ruprecht).

Gödel, K. 1947. In Benaceroff, P., and Putnam, H., eds. 1983. *Philosophy of Mathematics: Selected Readings*, 2nd ed. (Cambridge: Cambridge University Press).

Godwin, M., and Irvine, A. D. 2003. In Griffin, N., ed. *The Cambridge Companion to Bertrand Russell* (Cambridge: Cambridge University Press).

Goldstein, R. 2005. *Incompleteness: The Proof and Paradox of Kurt Gödel* (New York: W. W. Norton).

Gosling, J. C. B. 1973. *Plato* (London: Routledge & Kegan Paul).

Gott, J. R. 2001. *Time Travel in Einstein's Universe* (Boston: Houghton Mifflin).

Grassi, O. 1619. *Libra Astronomica ac Philosophica.* In Drake, S., and O'Malley, C. D., trans. 1960. *The Controversy on the Comets of 1618* (Philadelphia: University of Pennsylvania Press).

Graunt, J. 1662. *Natural and Political Observations Mentioned in a Following Index, and Made Upon the Bills of Mortality* (London: Tho. Roycroft).

Gray, J. J. 2004. *János Bolyai, Non-Euclidean Geometry, and the Nature of Space* (Cambridge, Mass.: Burndy Library).

Grayling, A. C. 2005. Descartes: *The Life and Times of a Genius* (New York: Walker & Company).

Greenberg, M. J. 1974. *Euclidean and Non-Euclidean Geometries: Development and History*, 3rd ed. (New York: W. H. Freeman and Company).

Greene, B. 1999. *The Elegant Universe: Superstrings, Hidden Dimensions, and the Quest for the Ultimate Theory* (New York: W. W. Norton).

——. 2004. *The Fabric of the Cosmos: Space, Time, and the Texture of Reality* (New York: Alfred A. Knopf).

Gross, D. 1988. *Proceedings of the National Academy of Sciences* (USA), 85, 8371.

Guthrie, K. S. 1987. *The Pythagorean Sourcebook and Library: An Anthology of Ancient Writings which Relate to Pythagoras and Pythagorean Philosophy* (Grand Rapids, Mich.: Phanes Press).

Hald, A. 1990. *A History of Probability and Statistics and Their Applications Before 1750* (New York: John Wiley & Sons).

Hall, A. R. 1992. *Isaac Newton: Adventurer in Thought* (Oxford: Blackwell). Reissued 1996 (Cambridge: Cambridge University Press).

Hamilton, E., and Cairns, H., eds. 1961. *The Collected Dialogues of Plato* (New York: Pantheon).

Hamming, R. W. 1980. *The American Mathematical Monthly,* 87(2), 81.

Hankins, F. H. 1908. *Adolphe Quetelet as Statistician* (New York: Columbia University). Posted online by R. E. Wyllys at http://www.gslis.utexas.edu/~wyllys/QueteletResources/index.html.

Hardy, G. H. 1940. *A Mathematician's Apology* (Cambridge: Cambridge University Press).

Havelock, E. 1963. *Preface to Plato* (Cambridge, Mass.: Harvard University Press).

Hawking, S. 2005. *God Created the Integers: The Mathematical Breakthroughs that Changed History* (Philadelphia: Running Press).

Hawking, S., ed. 2007. *A Stubbornly Persistent Illusion: The Essential Scientific Writings of Albert Einstein* (Philadelphia: Running Press).

Hawking, S., and Penrose, R. 1996. *The Nature of Space and Time* (Princeton: Princeton University Press).

Heath, T. L. 1897. *The Works of Archimedes* (Cambridge: Cambridge University Press).

——. 1921. *A History of Greek Mathematics* (Oxford: Clarendon Press). Republished 1981 (New York: Dover Publications).

Hedrick, P. W. 2004. *Genetics of Populations* (Sudbury, Mass.: Jones & Bartlett).

Heiberg, J. L., ed. 1910–15. *Archimedes Opera Omnio cum Commentariis Eutocii* (Leipzig); the text is in Greek with Latin translation.

Hellman, H. 2006. *Great Feuds in Mathematics: Ten of the Liveliest Disputes Ever* (Hoboken, N.J.: John Wiley & Sons).

Hermite, C. 1905. *Correspondence d'Hermite et de Stieltjes* (Paris: Gauthier-Villars).

Herodotus. 440 BC. *The History,* book III. Translated by D. Greve, 1988 (Chicago: University of Chicago Press).

Hersh, R. 2000. *18 Unconventional Essays on the Nature of Mathematics* (New York: Springer).

Herz-Fischler, R. 1998. *A Mathematical History of the Golden Number* (Mineola, N.Y.: Dover Publications).

Hobbes, T. 1651. *Leviathan.* Republished 1982 (New York: Penguin Classics).

Hockett, C. F. 1960. *Scientific American,* 203 (September), 88.

Höffe, O. 1994. *Immanuel Kant.* Translated by M. Farrier (Albany, N.Y.: SUNY Press).

Hofstadter, D. 1979. *Gödel, Escher, Bach: An Eternal Golden Braid* (New York: Basic Books).

Holden, C. 2006. *Science,* 311, 317.

Huffman, C. A. 1999. In Long, A. A., ed. *The Cambridge Companion to Early Greek Philosophy* (Cambridge: Cambridge University Press).

——. 2006. "Pythagoras." In the Stanford Encyclopedia of Philosophy. http://plato.stanford.edu/entries/pythagoras.

Hume, D. 1748. *An Enquiry Concerning Human Understanding.* Republished 2000 in *The Clarendon Edition of the Works of David Hume,* edited by T. L. Beauchamp (Oxford: Oxford University Press).

Iamblichus. Ca. 300 ADa. *Iamblichus' Life of Pythagoras.* Translated by T. Taylor, 1986 (Rochester, Vt.: Inner Traditions).

——. Ca. 300 ADb. *On the Pythagorean Life.* Translated by J. Dillon and J. Hershbell. (Atlanta: Scholar Press).

Irvine, A. D. 2003. "Russell's Paradox." In the Stanford Encyclopedia of Philosophy. http://plato.stanford.edu/entries/russell-paradox.

Isaacson, W. 2007. *Einstein: His Life and Universe* (New York: Simon & Schuster).

Jaeger, M. 2002. *The Journal of Roman Studies,* 92, 49.

Jeans, J. 1930. *The Mysterious Universe* (Cambridge: Cambridge University Press).

Jones, V. F. R. 1985. *Bulletin of the American Mathematical Society,* 12, 103.

Joost-Gaugier, C. L. 2006. *Measuring Heaven: Pythagoras and His Influence on Thought and Art in Antiquity and the Middle Ages* (Ithaca: Cornell University Press).

Kaku, M. 2004. *Einstein's Cosmos* (New York: Atlas Books).

Kant, I. 1781. *Critique of Pure Reason.* One of the many English translations is Müller, F. M. 1881. *Immanuel Kant's Critique of Pure Reason* (London: Macmillan).

Kaplan, M., and Kaplan, E. 2006. *Chances Are: Adventures in Probability* (New York: Viking).

Kapner, D. J., Cook, T. S., Adelberger, E. G., Gundlach, J. H., Heckel, B. R., Hoyle, C. D., and Swanson, H. E. 2007. *Physical Review Letters,* 98, 021101.

Kasner, E., and Newman, J. R. 1989. *Mathematics and the Imagination* (Redmond, Wash.: Tempus Books).

Kauffman, L. H. 2001. *Knots and Physics,* 3rd ed. (Singapore: World Scientific).

Keeling, S. V. 1968. *Descartes* (Oxford: Oxford University Press).

Kepler, J. 1981. *Mysterium Cosmographicum* (New York: Abaris Books).

——. 1997. *The Harmony of the World* (Philadelphia: American Philosophical Society).

Klessinger, N., Szczerbinski, M., and Varley, R. 2007. *Neuropsychologia,* 45, 1642.

Kline, M. 1967. *Mathematics for Liberal Arts* (Reading, Mass.: Addison-Wesley). Republished 1985 as *Mathematics for the Nonmathematician* (New York: Dover Publications).

——. 1972. *Mathematical Thought from Ancient to Modern Times* (Oxford: Oxford University Press).

Knott, C. G. 1911. *Life and Scientific Work of Peter Guthrie Tait* (Cambridge: Cambridge University Press).

Koyré, A. 1978. *Galileo Studies.* Translated by J. Mepham (Atlantic Highlands, N.J.: Humanities Press).

Kramer, M., Stairs, I. H., Manchester, R. N., et al. 2006. *Science,* 314 (5796), 97.

Krauss, L. 2005. *Hiding in the Mirror: The Mysterious Allure of Extra Dimensions, from Plato to String Theory and Beyond* (New York: Viking Penguin).

Kraut, R. 1992. *The Cambridge Companion to Plato* (Cambridge: Cambridge University Press).

Krüger, L. 1987. In Krüger, L., Daston, L. J., and Heidelberger, M., eds. *The Probabilistic Revolution* (Cambridge, Mass.: The MIT Press).

Kuehn, M. 2001. *Kant: A Biography* (Cambridge: Cambridge University Press).

Laertius, D. Ca. 250 AD. *Lives of Eminent Philosophers.* Translated by R. D. Hicks, 1925 (Cambridge, Mass.: Harvard University Press).

Lagrange, J. 1797. *Théorie des Fonctions Analytiques* (Paris: Imprimerie de la Republique).

Lahanas, M. "Archimedes and his Burning Mirrors." www.mlahanas.de/Greeks/Mirrors.htm.

Lakoff, G., and Núñez, R. E. 2000. *Where Mathematics Comes From* (New York: Basic Books).

Laplace, P. S., Marquis de. 1814. *A Philosophical Essay on Probabilities.* Translated by F. W. Truscot and F. L. Emory, 1902 (New York: John Wiley & Sons). Republished 1995 (Mineola, N.Y.: Dover).

Lecar, M., Franklin, F. A., Holman, M. J., and Murray, N. W. 2001. *Annual Review of Astronomy and Astrophysics*, 39, 581.

Lightman, A. 1993. *Einstein's Dreams* (New York: Pantheon Books).

Little, C. N. 1899. *Transaction of the Royal Society of Edinburgh*, 39 (part III), 771.

Livio, M. 2002. *The Golden Ratio: The Story of Phi, the World's Most Astonishing Number* (New York: Broadway Books).

——. 2005. *The Equation That Couldn't Be Solved* (New York: Simon & Schuster).

Lottin, J. 1912. *Quetelet: Staticien et Sociologue* (Louvain: Institut Supérieur de Philosophie).

MacHale, D. 1985. *George Boole: His Life and Work* (Dublin: Boole Press Limited).

Machamer, P. 1998. In Machamer, P., ed. *The Cambridge Companion to Galileo* (Cambridge: Cambridge University Press).

Manning, H. P. 1914. *Geometry of Four Dimensions* (London: Macmillan). Reprinted 1956 (New York: Dover Publications).

Maor, E. 1994. *e: The Story of a Number* (Princeton: Princeton University Press).

McMullin, E. 1998. In Machamer, P., ed. *The Cambridge Companion to Galileo* (Cambridge: Cambridge University Press).

Mekler, S., ed. 1902. *Academicorum Philosophorum Index Herculanensis* (Berlin: Weidmann).

Menasco, W., and Rudolph, L. 1995. *American Scientist*, 83 (January–February), 38.

Mendel, G. 1865. "Experiments in Plant Hybridization," http://www.mendelweb.org/Mendel.plain.html.

Merton, R. K. 1993. *On the Shoulders of Giants: A Shandean Postscript* (Chicago: University of Chicago Press).

Messer, R., and Straffin, P. 2006. *Topology Now* (Washington, D.C.: Mathematical Association of America).

Miller, V. R., and Miller, R. P., eds. 1983. *Descartes, Principles of Philosophy* (Dordrecht: Reidel).

Mitchell, J. C. 1990. In van Leeuwen, J., *Handbook of Theoretical Computer Science* (Cambridge, Mass.: MIT Press).

Monk, R. 1990. *Ludwig Wittgenstein: The Duty of Genius* (London: Jonathan Cape).

Moore, G. H. 1982. *Zermelo's Axiom of Choice: Its Origins, Development, and Influence* (New York: Springer-Verlag).

Morgenstern, O. 1971. Draft "Memorandum from Mathematica." Subject: History of the naturalization of Kurt Gödel. Institute for Advanced Study, Princeton, NJ.

Morris, T. 1999. *Philosophy for Dummies* (Foster City, Calif.: IDG Books).

Motte, A. 1729. *Sir Isaac Newton's Mathematical Principles of Natural Philosophy and His System of the World.* Revised by F. Cajori, 1947 (Berkeley: University of California Press). Also appeared as Newton, I. 1995. *The Principia* (New York: Prometheus Books).

Mueller, I. 1991. In Bowen, A., ed. *Science and Philosophy in Classical Greece* (London: Garland).

——. 1992. In Kraut, R., ed. *The Cambridge Companion to Plato* (Cambridge: Cambridge University Press).

——. 2005. In Koestier, T., and Bergmans, L., eds. *Mathematics and the Divine: A Historical Study* (Amsterdam: Elsevier).

Nagel, E., and Newman, J. 1959. *Gödel's Proof* (New York: Routledge & Kegan Paul). Republished 2001 (New York: New York University Press).

Netz, R. 2005. In Koetsier, T., and Bergmans, L., eds. *Mathematics and the Divine: A Historical Study* (Amsterdam: Elsevier).

Netz, R., and Noel, W. 2007. *The Archimedes Codex: How a Medieval Prayer Book Is Revealing the True Genius of Antiquity's Greatest Scientist* (Philadelphia: Da Capo Press).

Neuwirth, L. 1979. *Scientific American*, 240 (June), 110.

Newman, J. R. 1956. *The World of Mathematics* (New York: Simon & Schuster).

Newton, Sir I. 1729. *Mathematical Principles of Natural Philosophy.* Translated by I. B. Cohen and A. Whitman, 1999 (Berkeley: University of California Press).

——. 1730. *Opticks, or A Treatise of the Reflections, Refractions, Inflec-*

tions and Colours of Light, 4th ed. (London: G. Bell). Republished 1952 (New York: Dover Publications).
Nicolson, M. 1935. *Modern Philology*, 32(3), 233.
Obler, L. K., and Gjerlow, K. 1999. *Language and the Brain* (Cambridge: Cambridge University Press).
O'Connor, J. J., and Robertson, E. F. 2003. "Peter Guthrie Tait." http://www-history.mcs.st-andrews.ac.uk/Biographies/Tait.html.
——. 2005. "Hermann Günter Grassmann." http://www-history.mcs.st-andrews.ac.uk/Biographies/Grassmann.html.
——. 2007. "G. H. Hardy Addresses the British Association in 1922, part 1." http://www-history.mcs.st-andrews.ac.uk/Extras/BA_1922_1.html.
Odom, B., Hanneke, D., D'Urso, B., and Gabrielse, G. 2006. *Physical Review Letters*, 97, 030801.
Ooguri, H., and Vafa, C. 2000. *Nuclear Physics B*, 577, 419.
Orel, V. 1996. *Gregor Mendel: The First Geneticist* (New York: Oxford University Press).
Overbye, D. 2000. *Einstein in Love: A Scientific Romance* (New York: Viking).
Pais, A. 1982. *Subtle Is the Lord: The Science and Life of Albert Einstein* (Oxford: Oxford University Press).
Panek, R. 1998. *Seeing and Believing: How the Telescope Opened Our Eyes and Minds to the Heavens* (New York: Viking).
Paulos, J. A. 2008. *Irreligion: A Mathematician Explains Why the Arguments for God Just Don't Add Up* (New York: Hill and Wang).
Penrose, R. 1989. *The Emperor's New Mind: Concerning Computers, Minds, and the Laws of Physics* (Oxford: Oxford University Press).
——. 2004. *The Road to Reality: A Complete Guide to the Laws of the Universe* (London: Jonathan Cape).
Perko, K. A., Jr. 1974. *Proceedings of the American Mathematical Society*, 45, 262.
Pesic, P. 2007. *Beyond Geometry: Classic Papers from Riemann to Einstein* (Mineola, N.Y.: Dover Publications).
Peterson, I. 1988. *The Mathematical Tourist: Snapshots of Modern Mathematics* (New York: W. H. Freeman and Company).
Petsche, J.-J. 2006. *Grassmann* (Basel: Birkhäuser Verlag).
Pinker, S. 1994. *The Language Instinct* (New York: William Morrow and Company).
Plato. Ca. 360 BC. *The Republic.* Translated by A. Bloom, 1968 (New York: Basic Books).
Plutarch. Ca. 75 AD. "Marcellus." Translated by J. Dryden. In Clough, A. H., ed. 1992. *Plutarch's Lives* (New York: Modern Library).

Poincaré, H. 1891. *Revue Générale des Sciences Pures et Appliquées* 2, 769. The article is reprinted in English in Pesic, P., 2007. *Beyond Geometry.*

Porphyry. Ca. 270 AD. *Life of Pythagoras.* In Hadas, M., and Smith, M., eds. 1965. *Heroes and Gods* (New York: Harper and Row).

Proclus. Ca. 450. *Proclus: A Commentary on the First Book of Euclid's "Elements."* Translated by G. Morrow, 1970. (Princeton: Princeton University Press).

Przytycki, J. H. 1992. Aportaciones Matemáticas Comunicaciones, 11, 173.

Putnam, H. 1975. *Mathematics, Matter and Method: Philosophical Papers,* vol. 1 (Cambridge: Cambridge University Press), 60.

Quetelet, L. A. J. 1828. *Instructions Populaires sur le Calcul des Probabilités* (Brussels: H. Tarbier & M. Hayez).

Quine, W. V. O. 1966. *The Ways of Paradox and Other Essays* (New York: Random House).

——. 1982. *Methods of Logic,* 4th ed. (Cambridge, Mass.: Harvard University Press).

Radelet-de Grave, P., ed. 2005. "Bernoulli-Edition." http://www.ub.unibas.ch/spez/bernoull.htm.

Ramachandran, V. S., and Blakeslee, S. 1999. *Phantoms of the Brain* (New York: Quill).

Randall, L. 2005. *Warped Passages: Unraveling the Mysteries of the Universe's Hidden Dimensions* (New York: Ecco).

Raskin, J. 1998. "Effectiveness of Mathematics." http://jef.raskincenter.org/unpublished/effectiveness_mathematics .html.

Raymond, E. S. 2005. "The Utility of Mathematics." http://www.catb.org/~esr/writings/utility-of-math.

Redondi, P. 1998. In Machamer, P. *The Cambridge Companion to Galileo* (Cambridge: Cambridge University Press).

Rees, M. J. 1997. *Before the Beginning* (Reading, Mass.: Addison-Wesley).

Reeves, E. 2008. *Galileo's Glassworks: The Telescope and the Mirror* (Cambridge, Mass.: Harvard University Press).

Renon, L., and Felliozat, J. 1947. *L'Inde Classique: Manuel des Études Indiennes* (Paris: Payot).

Rescher, N. 2001. *Paradoxes: Their Roots, Range, and Resolution* (Chicago: Open Court).

Resnik, M. D. 1980. *Frege and the Philosophy of Mathematics* (Ithaca: Cornell University Press).

Reston, J. 1994. *Galileo: A Life* (New York: HarperCollins).

Ribenboim, P. 1994. *Catalan's Conjecture* (Boston: Academic Press).

Ricoeur, P. 1996. *Synthese,* 106, 57.

Riedweg, C. 2005. *Pythagoras: His Life and Influence.* Translated by S. Rendall (Ithaca: Cornell University Press).

Rivest, R., Shamir, A., and Adleman, L. 1978. *Communications of the Association for Computing Machinery*, 21(2), 120.

Rodis-Lewis, G. 1998. *Descartes: His Life and Thought* (Ithaca: Cornell University Press).

Ronan, M. 2006. *Symmetry and the Monster: The Story of One of the Greatest Quests of Mathematics* (New York: Oxford University Press).

Rosenthal, J. S. 2006. *Struck by Lightning: The Curious World of Probabilities* (Washington, D.C.: Joseph Henry Press).

Ross, W. D. 1951. *Plato's Theory of Ideas* (Oxford: Clarendon Press).

Rouse Ball, W. W. 1908. *A Short Account of the History of Mathematics*, 4th ed. Republished 1960 (Mineola, N.Y.: Dover Publications).

Rucker, R. 1995. *Infinity and the Mind: The Science and Philosophy of the Infinite* (Princeton: Princeton University Press).

Russell, B. 1912. *The Problems of Philosophy* (London: Home University Library). Reprinted 1997 by Oxford University Press (Oxford).

——. 1919. *Introduction to Mathematical Philosophy* (London: George Allen and Unwin). Reprinted 1993, edited by J. Slater (London: Routledge). Reprinted 2005 (New York: Barnes & Noble).

——. 1945. *History of Western Philosophy.* Reprinted 2007 (New York: Touchstone).

Sainsbury, R. M. 1988. *Paradoxes* (Cambridge: Cambridge University Press).

Sarrukai, S. 2005. *Current Science*, 88(3), 415.

Schmitt, C. B. 1969. "Experience and Experiment: A Comparison of Zabarella's Views with Galileo's in *De Motu.*" *Studies in the Renaissance*, 16, 80.

Sedgwick, W. T., and Tyler, H. W. 1917. *A Short History of Science* (New York: The Macmillan Company).

Shapiro, S. 2000. *Thinking about Mathematics: The Philosophy of Mathematics* (Oxford: Oxford University Press).

Shea, W. R. 1972. *Galileo's Intellectual Revolution: Middle Period, 1610–1632* (New York: Science History Publications).

——. 1998. In Machamer, P., ed. *The Cambridge Companion to Galileo* (Cambridge: Cambridge University Press).

Sieg, W. 1988. "Hilbert's Program Sixty Years Later." *Journal of Symbolic Logic*, 53, 349.

Smolin, L. 2001. *Three Roads to Quantum Gravity* (New York: Basic Books).

——. 2006. *The Trouble with Physics: The Rise of String Theory, The Fall of Science, and What Comes Next* (Boston: Houghton Mifflin).

Sobel, D. 1999. *Galileo's Daughter* (New York: Walker & Company).

Sommerville, D. M. Y. 1929. *An Introduction to the Geometry of N Dimensions* (London: Methuen).

Sorell, T. 2005. *Descartes Reinvented* (Cambridge: Cambridge University Press).

Sorensen, R. 2003. *A Brief History of the Paradox: Philosophy and the Labyrinths of the Mind* (Oxford: Oxford University Press).

Sossinsky, A. 2002. *Knots: Mathematics with a Twist* (Cambridge, Mass.: Harvard University Press).

Stanley, T. 1687. *The History of Philosophy*, ninth section. Published in 1970 as a photographic facsimile under the title *Pythagoras: His Life and Teachings* (Los Angeles: The Philosophical Research Society).

Steiner, M. 2005. In Shapiro, S., ed. *The Oxford Handbook of Philosophy of Mathematics and Logic* (Oxford: Oxford University Press).

Stewart, I. 2004. *Galois Theory* (Boca Raton, Fla.: Chapman & Hall/CRC).

——. 2007. *Why Beauty Is Truth: A History of Symmetry* (New York: Perseus Books).

Stewart, J. A. 1905. *The Myths of Plato* (London: Macmillan and Co.).

Stigler, S. M. 1997. In *Académie Royale de Belgique, Bulletin de la Classe des Sciences, Mémoires*, collection 8(3), 47.

Strohmeier, J., and Westbrook, P. 1999. *Divine Harmony* (Berkeley, Calif.: Berkeley Hills Books).

Stukeley, W. 1752. *Memoirs of Sir Isaac Newton's Life.* Reprinted 1936 (London: Taylor and Francis).

Summers, D. W. 1995. *Notices of the American Mathematical Society*, 42(5), 528.

Swerdlow, N. 1998. In Machamer, P., ed. *The Cambridge Companion to Galileo* (Cambridge: Cambridge University Press).

Tabak, J. 2004. *Probability and Statistics: The Science of Uncertainty* (New York: Facts on File).

Tait, P. G. 1898. In *Scientific Papers of Peter Guthrie Tait, vol. 1* (Cambridge: Cambridge University Press).

Tait, W. W. 1996. In Hart, W. D. *The Philosophy of Mathematics* (Oxford: Oxford University Press).

Tegmark, M. 2004. In Barrow, J. D., Davies, P. C. W., and Harper, C. L., Jr., eds. *Science and Ultimate Reality* (Cambridge: Cambridge University Press).

——. 2007a. "Shut Up and Calculate," arXiv 0709.4024 [hep-th].

——. 2007b. "The Mathematical Universe," arXiv 0704.0646 [gr-qc].

Tennant, N. 1997. *The Taming of the True* (Oxford: Oxford University Press).

Theon of Smyrna. Ca. 130 AD. *Mathematics, Useful for Understanding Plato.* Translated by R. Lawlor and D. Lawlor, 1979 (San Diego: Wizards Bookshelf).

Tiles, M. 1996. In Bunin, N., and Tsui-James, E. P., eds. *The Blackwell Companion to Philosophy* (Oxford: Blackwell Publishing).

Todhunter, I. 1865. *A History of the Mathematical Theory of Probability* (Cambridge: Macmillan and Co.).

Toffler, A. 1970. *Future Shock* (New York: Random House).

Trudeau, R. J. 1987. *The Non-Euclidean Revolution* (Boston: Birkhäuser).

Truesdell, C. 1960. *The Rotational Mechanics of Flexible or Elastic Bodies, 1638–1788, Leonhardi Euler Opera Omnia*, ser. II, vol. 11, part 2 (Zürich: Orell Füssli).

Turnbull, H. W., Scott, J. F., Hall, A. R., and Tilling, L., eds. 1959–77. *The Correspondence of Isaac Newton* (Cambridge: Cambridge University Press).

Urquhart, A. 2003. In Griffin, N., ed. *The Cambridge Companion to Bertrand Russell* (Cambridge: Cambridge University Press).

Vafa, C. 2000. In Arnold, V., Atiyah, M., Lax, P., and Mazur, B., eds. *Mathematics: Frontiers and Perspectives* (Providence, R.I.: American Mathematical Society).

Vandermonde, A. T. 1771. *L'Histoire de l'Académie des Sciences avec les Memoires* (Paris: Memoires de l'Academie Royale des Sciences).

Van der Waerden, B. L. 1983. *Geometry and Algebra in Ancient Civilizations* (Berlin: Springer-Verlag).

Van Heijenoort, J., ed. 1967. *From Frege to Gödel: A Source Book in Mathematical Logic* (Cambridge, Mass.: Harvard University Press).

Van Helden, A. 1996. *Proceedings of the American Philosophical Society*, 140, 358.

Van Helden, A., and Burr, E. 1995. The Galileo Project. http://galileo.rice.edu/index.html.

Van Stegt, W. P. 1998. In Mancosu, P., ed. *From Brouwer to Hilbert: The Debate on the Foundations of Mathematics in the 1920s* (Oxford: Oxford University Press).

Varley, R., Klessinger, N., Romanowski, C., and Siegal, M. 2005. *Proceedings of the National Academy of Sciences* (USA), 102, 3519.

Vawter, B. 1972. *Biblical Inspiration* (Philadelphia: Westminster).

Vilenkin, A. 2006. *Many Worlds in One: The Search for Other Universes* (New York: Hill and Wang).

Vitruvius, M. P. 1st century BC. *De Architectura.* In Rowland, I. D., and Howe, T. N., eds. 1999. *Ten Books on Architecture* (Cambridge: Cambridge University Press).

Vlostos, G. 1975. *Plato's Universe* (Seattle: University of Washington Press).

Von Gebler, K. 1879. *Galileo Galilei and the Roman Curia.* Translated by J. Sturge. Reprinted 1977 (Merrick, N.Y.: Richwood Publishing Company).

Vrooman, J. R. 1970. *René Descartes: A Biography* (New York: Putnam).

Waismann, F. 1979. *Ludwig Wittgenstein and the Vienna Circle: Conversations Recorded by Friedrich Waismann.* Edited by B. McGuinness; translated by J. Schulte and B. McGuinness (Oxford: Basel Blackwell).

Wallace, D. F. 2003. *Everything and More: A Compact History of Infinity* (New York: W. W. Norton).

Wallechinsky, D., and Wallace, I. 1975–81. "Biography of Scottish Child Prodigy Marjory Fleming, part 1." http://www.trivia-library.com/b/biography-of-scottish-child-prodigy-marjory-fleming-part-1.htm.

Wallis, J. 1685. *Treatise of Algebra.* Quoted in Manning, H. P. 1914. *Geometry of Four Dimensions* (London: Macmillan).

Wang, H. 1996. *A Logical Journey: From Gödel to Philosophy* (Cambridge, Mass.: MIT Press).

Washington, G. 1788. Letter to Nicholas Pike, June 20, 1788. In Fitzpatrick, J. C., ed. 1931–44. *Writings of George Washington* (Washington, D.C.: Government Printing Office). Quoted in Deutsch, K. L., and Nicgorski, W., eds. 1994. *Leo Strauss: Political Philosopher and Jewish Thinker* (Lanham, Md.: Rowman & Littlefield).

Wasserman, S. A., and Cozzarelli, N. R. 1986. *Science*, 232, 951.

Watson, R. 2002. *Cogito, Ergo Sum: The Life of René Descartes* (Boston: David R. Godine).

Weinberg, S. 1993. *Dreams of a Final Theory* (New York: Pantheon Books).

Wells, D. 1986. *The Penguin Dictionary of Curious and Interesting Numbers* (London: Penguin). Revised edition 1997.

Westfall, R. S. 1983. *Never at Rest: A Biography of Isaac Newton* (Cambridge: Cambridge University Press).

Whiston, W. 1753. *Memoirs of the Life and Writings of Mr. William Whiston, Containing, Memoirs of Several of His Friends Also*, 2nd ed. (London: Printed for J. Whiston and B. White).

White, L. A. 1947. *Philosophy of Science*, 14(4), 289.

White, N. P. 1992. In Kraut, R., ed. *The Cambridge Companion to Plato* (Cambridge: Cambridge University Press).

Whitehead, A. N. 1911. *An Introduction to Mathematics* (London: Williams & Norgate). Reprinted 1992 (Oxford: Oxford University Press).

——. 1929. *Process and Reality: An Essay in Cosmology.* Republished

1978, edited by D. R. Griffin and D. W. Sherburne (New York: Free Press).

Whitehead, A. N., and Russell, B. 1910. *Principia Mathematica* (Cambridge: Cambridge University Press). Second edition 1927.

Wigner, E. P. 1960. *Communications in Pure and Applied Mathematics*, vol. 13, no. 1. Reprinted in Saatz, T. L., and Weyl, F. J., eds. 1969. *The Spirit and the Uses of the Mathematical Sciences* (New York: McGraw-Hill).

Wilczek, F. 2006. *Physics Today*, 59 (November), 8.

——. 2007. *Physics Today*, 60 (May), 8.

Witten, E. 1989. *Communications in Mathematical Physics*, 121, 351.

Wolfram, S. 2002. *A New Kind of Science* (Champaign, Ill.: Wolfram Media).

Wolterstorff, N. 1999. In Sorell, T., ed. *Descartes* (Dartmouth: Ashgate).

Woodin, W. H. 2001a. *Notices of the American Mathematical Society*, 48(6), 567.

——. 2001b. *Notices of the American Mathematical Society*, 48(7), 681.

Wright, C. 1997. In Heck, R., ed. *Language, Thought, and Logic: Essays in Honour of Michael Dummett* (Oxford: Oxford University Press).

Zalta, E. N. 2005. "Gottlob Frege." *Stanford Encyclopedia of Philosophy*, http://plato.stanford.edu/entries/frege/.

——. 2007. "Frege's Logic, Theorem, and Foundations for Arithmetic." *Stanford Encyclopedia of Philosophy*, http://plato.stanford.edu/entries/frege-logic/.

Zweibach, B. A. 2004. *A First Course in String Theory* (Cambridge: Cambridge University Press).